Digital Minimalism

Praise for *Digital Minimalism*

'I challenge you not to devour this wonderful book in one sitting. I certainly did and I started applying Cal's ideas to my own life immediately' Greg McKeown, author of *Essentialism*

'You're not the user, you're the product. Hang up, log off, and tune in to a different way to be in the world. Bravo, Cal, smart advice for good people' Seth Godin, author of *This is Marketing*

'Cal Newport speaks human truth to digital power. He calls out our enslavement to modern devices and calmly presents a better way to live and work' Julia Hobsbawm, author of *Fully Connected*

'This book is an urgent call to action for anyone serious about being in command of their own life' Ryan Holiday, author of *The Obstacle is the Way*

'Cal Newport has discovered a cure for the techno-exhaustion that plagues our always-on, digitally-caffeinated culture' Joshua Fields Millburn, *The Minimalists*

'Cal Newport's *Digital Minimalism* is the best book I've read in some time about our fraught relationship with technology. If you're looking for a blueprint to guide you as you liberate yourself from the shackles of email, social networks, smartphones, and screens, let this book be your guide' Adam Alter, author of *Irresistible*

'I hope that everyone who owns a mobile phone and has been wondering where their time goes gets a chance to absorb the ideas in this book. Put more energy into what makes you happy, and ruthlessly strip away the things that don't' Pete Adeney, aka Mr. Money Mustache

'As a presence on the page, Newport is exceptional in the realm of self-help authors' *The New York Times Book Review*

Praise for *Deep Work*

'A compelling case for cultivating intense focus, and offers immediately actionable steps for infusing more of it into our lives' Adam Grant, author of *Originals* and *Deep Work*

'A wonderfully entangled, intertwined, and erudite series of strategies, philosophies, disciplines, and techniques to sharpen your focus and dive deep into your work' 800-CEO-READ on *Deep Work*

ABOUT THE AUTHOR

Cal Newport is a tenured professor of Computer Science at Georgetown University. In addition to his academic work, Newport writes about the intersection of these technologies with our personal and professional lives. He is the author of five books, including *So Good They Can't Ignore You* and the bestselling *Deep Work: Rules for Focused Success in a Distracted World*. Newport's ideas have been published in top print publications, including *The New York Times*, the *Wall Street Journal*, the *Economist*, the *Financial Times* and the *Guardian*, but as a dedicated digital minimalist, Newport has never had a social media account.

Digital
Minimalism

ON LIVING BETTER WITH
LESS TECHNOLOGY

Cal Newport

BUSINESS

PENGUIN BUSINESS

UK | USA | Canada | Ireland | Australia
India | New Zealand | South Africa

Penguin Business is part of the Penguin Random House group of companies
whose addresses can be found at global.penguinrandomhouse.com.

First published in the United States of America by Portfolio/Penguin,
an imprint of Penguin Random House LLC 2019
First published in Great Britain by Penguin Business 2019
002

Copyright © Calvin C. Newport, 2019

The moral right of the author has been asserted

Printed and bound in Great Britain by Clays Ltd, Elcograf S.p.A.

A CIP catalogue record for this book is available from the British Library

ISBN: 978-0-241-34113-1

To Julie:

my partner, my muse,

my voice of reason

Contents

Introduction

I n September 2016, the influential blogger and commentator
Andrew Sullivan wrote a 7,000-word essay for *New York*
magazine titled "I Used to Be a Human Being." Its subtitle
was alarming: "An endless bombardment of news and gossip
and images has rendered us manic information addicts. It
broke me. It might break you, too."

The article was widely shared. I'll admit, however, that
when I first read it, I didn't fully comprehend Sullivan's warning. I'm one of the few members of my generation to never
have a social media account, and tend not to spend much time
web surfing. As a result, my phone plays a relatively minor role
in my life—a fact that places me outside the mainstream experience this article addressed. In other words, I knew that the
innovations of the internet age were playing an increasingly
intrusive role in many people's lives, but I didn't have a *visceral*
understanding of what this meant. That is, until everything
changed.

Earlier in 2016, I published a book titled *Deep Work*. It was about the underappreciated value of intense focus and how the professional world's emphasis on distracting communication tools was holding people back from producing their best work. As my book found an audience, I began to hear from more and more of my readers. Some sent me messages, while others cornered me after public appearances—but many of them asked the same question: What about their personal lives? They agreed with my arguments about office distractions, but as they then explained, they were arguably even more distressed by the way new technologies seemed to be draining meaning and satisfaction from their time spent outside of work. This caught my attention and tumbled me unexpectedly into a crash course on the promises and perils of modern digital life.

Almost everyone I spoke to believed in the power of the internet, and recognized that it can and should be a force that improves their lives. They didn't necessarily want to give up Google Maps, or abandon Instagram, but they also felt as though their current relationship with technology was unsustainable—to the point that if something didn't change soon, they'd break, too.

A common term I heard in these conversations about modern digital life was *exhaustion*. It's not that any one app or website was particularly bad when considered in isolation. As many people clarified, the issue was the overall impact of having *so many* different shiny baubles pulling so insistently at their attention and manipulating their mood. Their problem with this frenzied activity is less about its details than the fact that it's increasingly beyond their control. Few want to spend so much time online, but these tools have a way of cultivating behavioral addictions. The urge to check Twitter or refresh Reddit becomes a nervous twitch that shatters uninterrupted time into shards too small to support the presence necessary for an intentional life.

As I discovered in my subsequent research, and will argue in the next chapter, some of these addictive properties are accidental (few predicted the extent to which text messaging could command your attention), while many are quite purposeful (compulsive use is the foundation for many social media business plans). But whatever its source, this irresistible attraction to screens is leading people to feel as though they're ceding more and more of their autonomy when it comes to deciding how they direct their attention. No one, of course, signed up for this loss of control. They downloaded the apps and set up accounts for good reasons, only to discover, with grim irony, that these services were beginning to undermine the very values that made them appealing in the first place: they joined Facebook to stay in touch with friends across the country, and then ended up unable to maintain an uninterrupted conversation with the friend sitting across the table.

I also learned about the negative impact of unrestricted online activity on psychological well-being. Many people I spoke to underscored social media's ability to manipulate their mood. The constant exposure to their friends' carefully curated portrayals of their lives generates feelings of inadequacy—especially during periods when they're already feeling low—and for teenagers, it provides a cruelly effective way to be publicly excluded.

In addition, as demonstrated during the 2016 presidential election and its aftermath, online discussion seems to accelerate people's shift toward emotionally charged and draining extremes. The techno-philosopher Jaron Lanier convincingly argues that the primacy of anger and outrage online is, in some sense, an unavoidable feature of the medium: In an open marketplace for attention, darker emotions attract more eyeballs than positive and constructive thoughts. For heavy internet users, repeated interaction with this darkness can become a source of draining negativity—a steep price that many don't even realize they're paying to support their compulsive connectivity.

Encountering this distressing collection of concerns—from the exhausting and addictive overuse of these tools, to their ability to reduce autonomy, decrease happiness, stoke darker instincts, and distract from more valuable activities—opened my eyes to the fraught relationship so many now maintain with the technologies that dominate our culture. It provided me, in other words, a much better understanding of what Andrew Sullivan meant when he lamented: "I used to be a human being."

■ ■ ■

This experience of talking with my readers convinced me that the impact of technology on people's personal lives was worth deeper exploration. I began more seriously researching and writing on this topic, trying to both better understand its contours and seek out the rare examples of those who can extract great value from these new technologies without losing control.*

One of the first things that became clear during this exploration is that our culture's relationship with these tools is complicated by the fact that they mix harm with benefits. Smartphones, ubiquitous wireless internet, digital platforms that connect billions of people—these are triumphant innovations! Few serious commentators think we'd be better off retreating to an earlier technological age. But at the same time, people are tired of feeling like they've become a slave to their devices. This reality creates a jumbled emotional landscape where you can simultaneously cherish your ability to discover inspiring photos on Instagram while fretting about this app's ability to invade the evening hours you used to spend talking with friends or reading.

The most common response to these complications is to suggest modest hacks and tips. Perhaps if you observe a digital Sabbath, or keep your phone away from your bed at night, or

---

* To some, the fact that I can't draw from a deep well of personal experience is a liability. "How can you criticize social media if you've never used it?" is one of the most common complaints I hear in response to my public advocacy on these issues. There's some truth to this claim, but as I recognized back in 2016 when I began this investigation, my outsider status can also prove advantageous. By approaching our tech culture from a fresh perspective, I'm perhaps better able to distinguish assumption from truth, and meaningful use from manipulation.

turn off notifications and resolve to be more mindful, you can keep all the good things that attracted you to these new technologies in the first place while still minimizing their worst impacts. I understand the appeal of this moderate approach because it relieves you of the need to make hard decisions about your digital life—you don't have to quit anything, miss out on any benefits, annoy any friends, or suffer any serious inconveniences.

But as is becoming increasingly clear to those who have attempted these types of minor corrections, willpower, tips, and vague resolutions are not sufficient by themselves to tame the ability of new technologies to invade your cognitive landscape—the addictiveness of their design and the strength of the cultural pressures supporting them are too strong for an ad hoc approach to succeed. In my work on this topic, I've become convinced that what you need instead is a full-fledged *philosophy of technology use*, rooted in your deep values, that provides clear answers to the questions of what tools you should use and how you should use them and, equally important, enables you to confidently ignore everything else.

There are many philosophies that might satisfy these goals. On one extreme, there are the Neo-Luddites, who advocate the abandonment of most new technologies. On another extreme, you have the Quantified Self enthusiasts, who carefully integrate digital devices into all aspects of their life with the goal of optimizing their existence. Of the different philosophies I studied, however, there was one in particular that stood out as a superior answer for those looking to thrive in our current moment of technological overload. I call it *digital minimalism*, and it applies the belief that *less can be more* to our relationship with digital tools.

This idea is not new. Long before Henry David Thoreau exclaimed "simplicity, simplicity, simplicity," Marcus Aurelius asked: "You see how few things you have to do to live a satisfying and reverent life?" Digital minimalism simply adapts this classical insight to the role of technology in our modern lives. The impact of this simple adaptation, however, can be profound. In this book, you'll encounter many examples of digital minimalists who experienced massively positive changes by ruthlessly reducing their time spent online to focus on a small number of high-value activities. Because digital minimalists spend so much less time connected than their peers, it's easy to think of their lifestyle as extreme, but the minimalists would argue that this perception is backward: what's extreme is how much time *everyone else* spends staring at their screens.

The key to thriving in our high-tech world, they've learned, is to spend much less time using technology.

■   ■   ■

The goal of this book is to make the case for digital minimalism, including a more detailed exploration of what it asks and why it works, and then to teach you how to adopt this philosophy if you decide it's right for you.

To do so, I divided the book into two parts. In part 1, I describe the philosophical underpinnings of digital minimalism, starting with a closer examination of the forces that are making so many people's digital lives increasingly intolerable, before moving on to a detailed discussion of the digital minimalism philosophy, including my argument for why it's the *right* solution to these problems.

Part 1 concludes by introducing my suggested method for adopting this philosophy: *the digital declutter.* As I've argued, aggressive action is needed to fundamentally transform your relationship with technology. The digital declutter provides this aggressive action.

This process requires you to step away from optional online activities for thirty days. During this period, you'll wean yourself from the cycles of addiction that many digital tools can instill, and begin to rediscover the analog activities that provide you deeper satisfaction. You'll take walks, talk to friends in person, engage your community, read books, and stare at the clouds. Most importantly, the declutter gives you the space to refine your understanding of the things you value most. At the end of the thirty days, you will then add back a small number of carefully chosen online activities that you believe will provide massive benefit to these things you value. Going forward, you'll do your best to make these intentional activities the core of your online life—leaving behind most of the other distracting behaviors that used to fragment your time and snare your attention. The declutter acts as a jarring reset: you come into the process a frazzled maximalist and leave an intentional minimalist.

In this final chapter of part 1, I'll guide you through implementing your own digital declutter. In doing so, I'll draw extensively on an experiment I ran in the early winter of 2018 in which over 1,600 people agreed to perform a digital declutter under my guidance and report back about their experience. You'll hear these participants' stories and learn what strategies worked well for them, and what traps they encountered that you should avoid.

The second part of this book takes a closer look at some ideas that will help you cultivate a sustainable digital minimalism lifestyle. In these chapters, I examine issues such as the importance of solitude and the necessity of cultivating high-quality leisure to replace the time most now dedicate to mindless device use. I propose and defend the perhaps controversial claim that your relationships will *strengthen* if you stop clicking "Like" or leaving comments on social media posts, and become harder to reach by text messages. I also provide an insider look at the *attention resistance*—a loosely organized movement of individuals who use high-tech tools and strict operating procedures to extract value from the products of the digital attention economy, while avoiding falling victim to compulsive use.

Each chapter in part 2 concludes with a collection of *practices*, which are concrete tactics designed to help you act on the big ideas of the chapter. As a budding digital minimalist, you can view the part 2 practices as a toolbox meant to aid your efforts to build a minimalist lifestyle that works for your particular circumstances.

■ ■ ■

In *Walden*, Thoreau famously writes: "The mass of men lead lives of quiet desperation." Less often quoted, however, is the optimistic rejoinder that follows in his next paragraph:

> They honestly think there is no choice left. But alert
> and healthy natures remember that the sun rose clear.
> It is never too late to give up our prejudices.

Our current relationship with the technologies of our hyper-connected world is unsustainable and is leading us closer to the quiet desperation that Thoreau observed so many years ago. But as Thoreau reminds us, "the sun rose clear" and we still have the ability to change this state of affairs.

To do so, however, we cannot passively allow the wild tangle of tools, entertainments, and distractions provided by the internet age to dictate how we spend our time or how we feel. We must instead take steps to extract the good from these technologies while sidestepping what's bad. We require a philosophy that puts our aspirations and values once again in charge of our daily experience, all the while dethroning primal whims and the business models of Silicon Valley from their current dominance of this role; a philosophy that accepts new technologies, but not if the price is the dehumanization Andrew Sullivan warned us about; a philosophy that prioritizes long-term meaning over short-term satisfaction.

A philosophy, in other words, like digital minimalism.

PART 1

Foundations

1

A Lopsided Arms Race

WE DIDN'T SIGN UP FOR THIS

I remember when I first encountered Facebook: It was the spring of 2004; I was a senior in college and began to notice an increasing number of my friends talk about a website called thefacebook.com. The first person to show me an actual Facebook profile was Julie, who was then my girlfriend, and now my wife.

"My memory of it was that it was a novelty," she told me recently. "It had been sold to us as a virtual version of our printed freshman directory, something we could use to look up the boyfriends or girlfriends of people we knew."

The key word in this memory is *novelty*. Facebook didn't arrive in our world with a promise to radically transform the rhythms of our social and civic lives; it was just one diversion among many. In the spring of 2004, the people I knew who signed up for thefacebook.com were almost certainly spending significantly more time playing Snood (a Tetris-style puzzle game that was inexplicably popular) than they were tweaking their profiles or poking their virtual friends.

"It was interesting," Julie summarized, "but it certainly didn't seem like this was something on which we would spend any real amount of time."

Three years later, Apple released the iPhone, sparking the mobile revolution. What many forget, however, was that the original "revolution" promised by this device was also much more modest than the impact it eventually created. In our current moment, smartphones have reshaped people's experience of the world by providing an always-present connection to a humming matrix of chatter and distraction. In January 2007, when Steve Jobs revealed the iPhone during his famous Macworld keynote, the vision was much less grandiose.

One of the major selling points of the original iPhone was that it integrated your iPod with your cell phone, preventing you from having to carry around two separate devices in your pockets. (This is certainly how I remember thinking about the iPhone's benefits when it was first announced.) Accordingly, when Jobs demonstrated an iPhone onstage during his keynote address, he spent the first eight minutes of the demo walking through its media features, concluding: "It's the best iPod we've ever made!"

Another major selling point of the device when it launched was the many ways in which it improved the experience of making phone calls. It was big news at the time that Apple forced AT&T to open its voicemail system to enable a better

interface for the iPhone. Onstage, Jobs was also clearly enam-
ored of the simplicity with which you could scroll through
phone numbers, and the fact that the dial pad appeared on the
screen instead of requiring permanent plastic buttons.

"The killer app is making calls," Jobs exclaimed to applause
during his keynote. It's not until thirty-three minutes into
that famed presentation that he gets around to highlighting
features like improved text messaging and mobile internet access that dominate the way we now use these devices.

To confirm that this limited vision was not some quirk of
Jobs's keynote script, I spoke with Andy Grignon, who was
one of the original iPhone team members. "This was supposed
to be an iPod that made phone calls," he confirmed. "Our core
mission was playing music and making phone calls." As Grignon then explained to me, Steve Jobs was initially dismissive
of the idea that the iPhone would become more of a general-
purpose mobile computer running a variety of different third-
party applications. "The second we allow some knucklehead
programmer to write some code that crashes it," Jobs once told
Grignon, "that will be when they want to call 911."

When the iPhone first shipped in 2007, there was no App
Store, no social media notifications, no quick snapping of photos to Instagram, no reason to surreptitiously glance down a
dozen times during a dinner—and this was absolutely fine
with Steve Jobs, and the millions who bought their first smartphone during this period. As with the early Facebook adopt-
ers, few predicted how much our relationship with this shiny
new tool would mutate in the years that followed.

■ ■ ■

It's widely accepted that new technologies such as social media and smartphones massively changed how we live in the twenty-first century. There are many ways to portray this change. I think the social critic Laurence Scott does so quite effectively when he describes the modern hyper-connected existence as one in which "a moment can feel strangely flat if it exists solely in itself."

The point of the above observations, however, is to emphasize what many also forget, which is that these changes, in addition to being massive and transformational, were also unexpected and unplanned. A college senior who set up an account on thefacebook.com in 2004 to look up classmates probably didn't predict that the average modern user would spend around two hours *per day* on social media and related messaging services, with close to half that time dedicated to Facebook's products alone. Similarly, a first adopter who picked up an iPhone in 2007 for the music features would be less enthusiastic if told that within a decade he could expect to compulsively check the device eighty-five times a day—a "feature" we now know Steve Jobs never considered as he prepared his famous keynote.

These changes crept up on us and happened fast, before we had a chance to step back and ask what *we really wanted* out of the rapid advances of the past decade. We added new technologies to the periphery of our experience for minor reasons, then woke one morning to discover that they had colonized the core of our daily life. We didn't, in other words, sign up for the digital world in which we're currently entrenched; we seem to have stumbled backward into it.

This nuance is often missed in our cultural conversation surrounding these tools. In my experience, when concerns about new technologies are publicly discussed, techno-apologists are quick to push back by turning the discussion to utility— providing case studies, for example, of a struggling artist finding an audience through social media,* or WhatsApp connecting a deployed soldier with her family back home. They then conclude that it's incorrect to dismiss these technologies on the grounds that they're useless, a tactic that is usually sufficient to end the debate.

The techno-apologists are right in their claims, but they're also missing the point. The perceived utility of these tools is not the ground on which our growing wariness builds. If you ask the average social media user, for example, why they use Facebook, or Instagram, or Twitter, they can provide you with reasonable answers. Each one of these services probably offers them something useful that would be hard to find elsewhere:

---

* This example comes from personal experience. In the fall of 2016, I appeared on a national radio show on the CBC network in Canada to discuss a *New York Times* column I wrote questioning the benefits of social media for career advancement. The host surprised me early in the interview by bringing into the discussion an unannounced guest: an artist who promotes his work through social media. Funnily enough, not long into the interview, the artist admitted (unprompted) that he was finding social media to be too distracting and that he now takes long breaks from it to get work done.

the ability, for example, to keep up with baby pictures of a sibling's child, or to use a hashtag to monitor a grassroots movement.

The source of our unease is not evident in these thin-sliced case studies, but instead becomes visible only when confronting the thicker reality of how these technologies as a whole have managed to expand beyond the minor roles for which we initially adopted them. Increasingly, they dictate how we behave and how we feel, and somehow coerce us to use them more than we think is healthy, often at the expense of other activities we find more valuable. What's making us uncomfortable, in other words, is this feeling of *losing control*—a feeling that instantiates itself in a dozen different ways each day, such as when we tune out with our phone during our child's bath time, or lose our ability to enjoy a nice moment without a frantic urge to document it for a virtual audience.

It's not about usefulness, it's about autonomy.

The obvious next question, of course, is how we got ourselves into this mess. In my experience, most people who struggle with the online part of their lives are not weak willed or stupid. They're instead successful professionals, striving students, loving parents; they are organized and used to pursuing hard goals. Yet somehow the apps and sites beckoning from behind the phone and tablet screen—unique among the many temptations they successfully resist daily—managed to succeed in metastasizing unhealthily far beyond their original roles.

A large part of the answer about how this happened is that

many of these new tools are not nearly as innocent as they might first seem. People don't succumb to screens because they're lazy, but instead because billions of dollars have been invested to make this outcome inevitable. Earlier I noted that we seem to have stumbled backward into a digital life we didn't sign up for. As I'll argue next, it's probably more accurate to say that we were *pushed* into it by the high-end device companies and attention economy conglomerates who discovered there are vast fortunes to be made in a culture dominated by gadgets and apps.

TOBACCO FARMERS IN T-SHIRTS

Bill Maher ends every episode of his HBO show *Real Time* with a monologue. The topics are usually political. This was not the case, however, on May 12, 2017, when Maher looked into the camera and said:

> The tycoons of social media have to stop pretending that they're friendly nerd gods building a better world and admit they're just tobacco farmers in T-shirts selling an addictive product to children. Because, let's face it, checking your "likes" is the new smoking.

Maher's concern with social media was sparked by a *60 Minutes* segment that aired a month earlier. The segment is titled "Brain Hacking," and it opens with Anderson Cooper interviewing a lean, red-haired engineer with the carefully tended stubble popular among young men in Silicon Valley. His name is Tristan Harris, a former start-up founder and Google engineer who deviated from his well-worn path through the world of tech to become something decidedly rarer in this closed world: a whistleblower.

"This thing is a slot machine," Harris says early in the interview while holding up his smartphone.

"How is that a slot machine?" Cooper asks.

"Well, every time I check my phone, I'm playing the slot machine to see 'What did I get?'" Harris answers. "There's a whole playbook of techniques that get used [by technology companies] to get you using the product for as long as possible."

"Is Silicon Valley programming apps or are they programming people?" Cooper asks.

"They are programming people," Harris says. "There's always this narrative that technology's neutral. And it's up to us to choose how we use it. This is just not true—"

"Technology is not neutral?" Cooper interrupts.

"It's not neutral. They want you to use it in particular ways and for long periods of time. Because that's how they make their money."

Bill Maher, for his part, thought this interview seemed familiar. After playing a clip of the Harris interview for his HBO audience, Maher quips: "Where have I heard this before?" He then cuts to Mike Wallace's famous 1995 interview with Jeffrey Wigand—the whistleblower who confirmed for the world what most already suspected: that the big tobacco companies engineered cigarettes to be more addictive.

"Philip Morris just wanted your lungs," Maher concludes. "The App Store wants your soul."

■ ■ ■

Harris's transformation into a whistleblower is exceptional in part because his life leading up to it was so normal by Silicon Valley standards. Harris, who at the time of this writing is in his midthirties, was raised in the Bay Area. Like many engineers, he grew up hacking his Macintosh and writing computer code. He went to Stanford to study computer science and, after graduating, started a master's degree working in BJ Fogg's famed Persuasive Technology Lab—which explores how to use technology to change how people think and act. In Silicon Valley, Fogg is known as the "millionaire maker," a reference to the many people who passed through his lab and then applied what they learned to help build lucrative tech start-ups (a group that includes, among other dot-com luminaries, Instagram co-founder Mike Krieger). Following this established path, Harris, once sufficiently schooled in the art of mind-device interaction, dropped out of the master's program to found Apture, a tech start-up that used pop-up factoids to increase the time users spent on websites.

In 2011, Google acquired Apture, and Harris was put to work on the Gmail inbox team. It was at Google where Harris, now working on products that could impact hundreds of millions of people's behaviors, began to grow concerned. After a mind-opening experience at Burning Man, Harris, in a move straight out of a Cameron Crowe screenplay, wrote a 144-slide manifesto titled "A Call to Minimize Distraction & Respect Users' Attention." Harris sent the manifesto to a small group of friends at Google. It soon spread to thousands in the company, including co-CEO Larry Page, who called Harris into a meeting to discuss the bold ideas. Page named Harris to the newly invented position of "product philosopher."

But then: Nothing much changed. In a 2016 profile in the *Atlantic*, Harris blamed the lack of changes to the "inertia" of the organization and a lack of clarity about what he was advocating. The primary source of friction, of course, is almost certainly more simple: Minimizing distraction and respecting users' attention would reduce revenue. Compulsive use sells, which Harris now acknowledges when he claims that the attention economy drives companies like Google into a "race to the bottom of the brain stem."

So Harris quit, started a nonprofit called Time Well Spent with the mission of demanding technology that "serves us, not advertising," and went public with his warnings about how far technology companies are going to try to "hijack" our minds.

In Washington, DC, where I live, it's well-known that the biggest political scandals are those that confirm a negative that most people already suspected to be true. This insight perhaps explains the fervor that greeted Harris's revelations. Soon after going public, he was featured on the cover of the *Atlantic*, interviewed on *60 Minutes* and *PBS NewsHour*, and was whisked off to give a TED talk. For years, those of us who were grumbling about the seeming ease with which people were becoming slaves to their smartphones were put down as alarmist. But then Harris came along and confirmed what many were increasingly suspecting to be true: These apps and slick sites were not, as Bill Maher put it, gifts from "nerd gods building a better world." They were, instead, designed to put slot machines in our pockets.

Harris had the moral courage to warn us about the hidden dangers of our devices. If we want to thwart their worst effects, however, we need to better understand how they're so easily able to subvert our best intentions for our lives. Fortunately, when it comes to this goal, we have a good guide. As it turns out, during the same years when Harris was wrestling with the ethical impact of addictive technology, a young marketing professor at NYU turned his prodigious focus to figuring out how exactly this techno-addiction works.

■ ■ ■

Before 2013, Adam Alter had little interest in technology as a research subject. A business professor with a PhD from Princeton in social psychology, Alter studied the broad question of how features in the world around us influence our thoughts and behavior.

Alter's doctoral dissertation, for example, studies how coincidental connections between you and another person can impact how you feel about each other. "If you find out you have the same birthday as someone who does something horrible," Alter explained to me, "you hate them even more than if you didn't have that information."

His first book, *Drunk Tank Pink*, cataloged numerous similar cases where seemingly small environmental factors create large changes in behavior. The title, for example, refers to a study that showed aggressively drunk inmates at a Seattle naval prison were notably calmed after spending just fifteen minutes in a cell painted a particular shade of Pepto-Bismol pink, as were Canadian schoolchildren when taught in a classroom of the same color. The book also reveals that wearing a red shirt on a dating profile will lead to significantly more interest than any other color, and that the easier your name is to pronounce, the faster you'll advance in the legal profession.

What made 2013 a turning point for Alter's career was a cross-country flight from New York to LA. "I had grand plans to get some sleep and do some work," he told me. "But as we started taxiing to take off, I began playing a simple strategy game on my phone called 2048. When we landed six hours later, I was still playing the game."

After publishing *Drunk Tank Pink*, Alter had begun searching for a new topic to pursue—a quest that kept leading him back to a key question: "What's the single biggest factor shaping our lives today?" His experience of compulsive game playing on his six-hour flight suddenly snapped the answer into sharp focus: *our screens.*

By this point, of course, others had already started asking critical questions about our seemingly unhealthy relationship with new technologies like smartphones and video games, but what set Alter apart was his training in psychology. Instead of approaching the issue as a cultural phenomenon, he focused on its psychological roots. This new perspective led Alter inevitably and unambiguously in an unnerving direction: the science of addiction.

■ ■ ■

To many people, *addiction* is a scary word. In popular culture, it conjures images of drug addicts stealing their mother's jewelry. But to psychologists, addiction has a careful definition that's stripped of these more lurid elements. Here's a representative example:

> Addiction is a condition in which a person engages in use of a substance or in a behavior for which the rewarding effects provide a compelling incentive to repeatedly pursue the behavior despite detrimental consequences.

Until recently, it was assumed that addiction only applied to alcohol or drugs: substances that include psychoactive compounds that can directly change your brain chemistry. As the twentieth century gave way to the twenty-first, however, a mounting body of research suggested that behaviors that did not involve ingesting substances could become addictive in the technical sense defined above. An important 2010 survey paper, for example, appearing in the *American Journal of Drug and Alcohol Abuse,* concluded that "growing evidence suggests that behavioral addictions resemble substance addictions in many domains." The article points to pathological gambling and internet addiction as two particularly well-established examples of these disorders. When the American Psychiatric Association published its fifth edition of the *Diagnostic and Statistical Manual of Mental Disorders (DSM-5)* in 2013, it included, for the first time, behavioral addiction as a diagnosable problem.

This brings us back to Adam Alter. After reviewing the relevant psychology literature and interviewing relevant people in the technology world, two things became clear to him. First, our new technologies are particularly well suited to foster behavioral addictions. As Alter admits, the behavioral addictions connected to technology tend to be "moderate" as compared to the strong chemical dependencies created by drugs and cigarettes. If I force you to quit Facebook, you're not likely to suffer serious withdrawal symptoms or sneak out in the night to an internet café to get a fix. On the other hand, these addictions can still be quite harmful to your well-being. You might not sneak out to access Facebook, but if the app is only one tap away on the phone in your pocket, a moderate behavioral addiction will make it really hard to resist checking your account again and again throughout the day.

The second thing that became clear to Alter during his research is even more disturbing. Just as Tristan Harris warned, in many cases these addictive properties of new technologies are not accidents, but instead carefully engineered design features.

The natural follow-up question to Alter's conclusions is: What specifically makes new technologies well suited to foster behavioral addictions? In his 2017 book, *Irresistible*, which details his study of this topic, Alter explores the many different "ingredients" that make a given technology likely to hook our brain and cultivate unhealthy use. I want to briefly focus on two forces from this longer treatment that not only seemed particularly relevant to our discussion, but as you'll soon learn, repeatedly came up in my own research on how tech companies encourage behavioral addiction: *intermittent positive reinforcement* and *the drive for social approval*.

Our brains are highly susceptible to these forces. This matters because many of the apps and sites that keep people compulsively checking their smartphones and opening browser tabs often leverage these hooks to make themselves nearly impossible to resist. To understand this claim, let's briefly discuss both.

■ ■ ■

We begin with the first force: intermittent positive reinforcement. Scientists have known since Michael Zeiler's famous pecking pigeon experiments from the 1970s that rewards delivered unpredictably are far more enticing than those delivered with a known pattern. Something about unpredictability releases more dopamine—a key neurotransmitter for regulating our sense of craving. The original Zeiler experiment had pigeons pecking a button that unpredictably released a food pellet. As Adam Alter points out, this same basic behavior is replicated in the feedback buttons that have accompanied most social media posts since Facebook introduced the "Like" icon in 2009.

"It's hard to exaggerate how much the 'like' button changed the psychology of Facebook use," Alter writes. "What had begun as a passive way to track your friends' lives was now deeply interactive, and with exactly the sort of unpredictable feedback that motivated Zeiler's pigeons." Alter goes on to describe users as "gambling" every time they post something on a social media platform: Will you get likes (or hearts or retweets), or will it languish with no feedback? The former creates what one Facebook engineer calls "bright dings of pseudo-pleasure," while the latter feels bad. Either way, the outcome is hard to predict, which, as the psychology of addiction teaches us, makes the whole activity of posting and checking maddeningly appealing.

Social media feedback, however, is not the only online activity with this property of unpredictable reinforcement. Many people have the experience of visiting a content website for a specific purpose—say, for example, going to a newspaper site to check the weather forecast—and then find themselves thirty minutes later still mindlessly following trails of links, skipping from one headline to another. This behavior can also be sparked by unpredictable feedback: most articles end up duds, but occasionally you'll land on one that creates a strong emotion, be it righteous anger or laughter. Every appealing headline clicked or intriguing link tabbed is another meta-phorical pull of the slot machine handle.

Technology companies, of course, recognize the power of this unpredictable positive feedback hook and tweak their products with it in mind to make their appeal even stronger. As whistleblower Tristan Harris explains: "Apps and websites sprinkle intermittent variable rewards all over their products because it's good for business." Attention-catching notification badges, or the satisfying way a single finger swipe swoops in the next potentially interesting post, are often carefully tailored to elicit strong responses. As Harris notes, the notification symbol for Facebook was originally blue, to match the palette of the rest of the site, "but no one used it." So they changed the color to red—an alarm color—and clicking skyrocketed.

In perhaps the most telling admission of all, in the fall of 2017, Sean Parker, the founding president of Facebook, spoke candidly at an event about the attention engineering deployed by his former company:

> The thought process that went into building these applications, Facebook being the first of them, . . . was all about: "How do we consume as much of your time and conscious attention as possible?" And that means that we need to sort of give you a little dopamine hit every once in a while, because someone liked or commented on a photo or a post or whatever.

The whole social media dynamic of posting content, and then watching feedback trickle back unpredictably, seems fundamental to these services, but as Tristan Harris points out, it's actually just one arbitrary option among many for how they could operate. Remember that early social media sites featured very little feedback—their operations focused instead on posting and finding information. It tends to be these early, pre-feedback-era features that people cite when explaining why social media is important to their life. When justifying Facebook use, for example, many will point to something like the ability to find out when a friend's new baby is born, which is a one-way transfer of information that does not require feedback (it's implied that people "like" this news).

In other words, there's nothing fundamental about the unpredictable feedback that dominates most social media services. If you took these features away, you probably wouldn't diminish the value most people derive from them. The reason this specific dynamic is so universal is because it works really well for keeping eyes glued to screens. These powerful psychological forces are a large part of what Harris had in mind when he held up a smartphone on *60 Minutes* and told Anderson Cooper "this thing is a slot machine."

■ ■ ■

Let's now consider the second force that encourages behavioral addiction: the drive for social approval. As Adam Alter writes: "We're social beings who can't ever completely ignore what other people think of us." This behavior, of course, is adaptive. In Paleolithic times, it was important that you care-fully managed your social standing with other members of your tribe because your survival depended on it. In the twenty-first century, however, new technologies have hijacked this deep drive to create profitable behavioral addictions.

Consider, once again, social media feedback buttons. In addition to delivering unpredictable feedback, as discussed above, this feedback also concerns other people's approval. If lots of people click the little heart icon under your latest Ins-tagram post, it feels like the tribe is showing you approval—which we're adapted to strongly crave.* The other side of this evolutionary bargain, of course, is that a lack of positive feed-back creates a sense of distress. This is serious business for the Paleolithic brain, and therefore it can develop an urgent need to continually monitor this "vital" information.

The power of this drive for social approval should not be underestimated. Leah Pearlman, who was a product manager on the team that developed the "Like" button for Facebook (she was the author of the blog post announcing the feature in 2009), has become so wary of the havoc it causes that now, as a small business owner, she hires a social media manager to handle her Facebook account so she can avoid exposure to the service's manipulation of the human social drive. "Whether there's a notification or not, it doesn't really feel that good,"

* For a good introduction to the evolution of "groupish" instincts in human beings and their central role in how we make sense of the world, see Jonathan Haidt's illuminating book *The Righteous Mind* (New York: Pantheon, 2012).

Pearlman said about the experience of checking social media feedback. "Whatever we're hoping to see, it never quite meets that bar."

A similar drive to regulate social approval helps explain the current obsession among teenagers to maintain Snapchat "streaks" with their friends, as a long unbroken streak of daily communication is a satisfying confirmation that the relationship is strong. It also explains the universal urge to immediately answer an incoming text, even in the most inappropriate or dangerous conditions (think: behind the wheel). Our Paleolithic brain categorizes ignoring a newly arrived text the same as snubbing the tribe member trying to attract your attention by the communal fire: a potentially dangerous social faux pas.

The technology industry has become adept at exploiting this instinct for approval. Social media, in particular, is now carefully tuned to offer you a rich stream of information about how much (or how little) your friends are thinking about you at the moment. Tristan Harris highlights the example of tagging people in photos on services like Facebook, Snapchat, and Instagram. When you post a photo using these services, you can "tag" the other users who also appear in the photo. This tagging process sends the target of the tag a notification. As Harris explains, these services now make this process near automatic by using cutting-edge image recognition algorithms to figure out who is in your photos and offer you the ability to tag them with just a single click—an offer usually made in the form of a quick yes/no question ("do you want to tag . . . ?") to which you'll almost certainly answer yes.

This single click requires almost no effort on your part, but to the user being tagged, the resulting notification creates a socially satisfying sense that *you were thinking about them*. As Harris argues, these companies didn't invest the massive resources necessary to perfect this auto-tagging feature because it was somehow crucial to their social network's usefulness. They instead made this investment so they could significantly increase the amount of addictive nuggets of social approval that their apps could deliver to their users.

As Sean Parker confirmed in describing the design philosophy behind these features: "It's a social-validation feedback loop . . . exactly the kind of thing that a hacker like myself would come up with, because you're exploiting a vulnerability in human psychology."

■ ■ ■

Let's step back for a moment to review where we stand. In the preceding sections, I detailed a distressing explanation for why so many people feel as though they've lost control of their digital lives: the hot new technologies that emerged in the past decade or so are particularly well suited to foster behavioral addictions, leading people to use them much more than they think is useful or healthy. Indeed, as revealed by whistleblowers and researchers like Tristan Harris, Sean Parker, Leah Pearlman, and Adam Alter, these technologies are in many cases *specifically* designed to trigger this addictive behavior. Compulsive use, in this context, is not the result of a character flaw, but instead the realization of a massively profitable business plan.

We didn't sign up for the digital lives we now lead. They were instead, to a large extent, crafted in boardrooms to serve the interests of a select group of technology investors.

A LOPSIDED ARMS RACE

As argued, our current unease with new technologies is not really about whether or not they're useful. It's instead about autonomy. We signed up for these services and bought these devices for minor reasons—to look up friends' relationship statuses or eliminate the need to carry a separate iPod and phone—and then found ourselves, years later, increasingly dominated by their influence, allowing them to control more and more of how we spend our time, how we feel, and how we behave.

The fact that our humanity was routed by these tools over the past decade should come as no surprise. As I just detailed, we've been engaging in a lopsided arms race in which the technologies encroaching on our autonomy were preying with increasing precision on deep-seated vulnerabilities in our brains, while we still naively believed that we were just fiddling with fun gifts handed down from the nerd gods.

When Bill Maher joked that the App Store was coming for

our souls, he was actually onto something. As Socrates explained to Phaedrus in Plato's famous chariot metaphor, our
soul can be understood as a chariot driver struggling to rein
two horses, one representing our better nature and the other
our baser impulses. When we increasingly cede autonomy to
the digital, we energize the latter horse and make the chariot
driver's struggle to steer increasingly difficult—a diminishing
of our soul's authority.

When seen from this perspective, it becomes clear that this
is a battle we must fight. But to do so, we need a more serious
strategy, something custom built to swat aside the forces manipulating us toward behavioral addictions and that offers a
concrete plan about how to put new technologies to use *for* our
best aspirations and not *against* them. Digital minimalism is
one such strategy. It's toward its details that we now turn our
attention.

2

Digital Minimalism

A MINIMAL SOLUTION

Around the time I started working on this chapter, a columnist for the *New York Post* published an op-ed titled "How I Kicked the Smartphone Addiction—and You Can Too." His secret? He disabled notifications for 112 different apps on his iPhone. "It's relatively easy to retake control," he optimistically concludes.

These types of articles are common in the world of technology journalism. The author discovers that his relationship with his digital tools has become dysfunctional. Alarmed, he deploys a clever life hack, then reports enthusiastically that things seem much better. I'm always skeptical about these quick-fix tales. In my experience covering these topics, it's hard to permanently reform your digital life through the use of tips and tricks alone.

The problem is that small changes are not enough to solve

our big issues with new technologies. The underlying behaviors we hope to fix are ingrained in our culture, and, as I argued in the previous chapter, they're backed by powerful psychological forces that empower our base instincts. To reestablish control, we need to move beyond tweaks and instead rebuild our relationship with technology from scratch, using our deeply held values as a foundation.

The *New York Post* columnist cited above, in other words, should look beyond the notification settings on his 112 apps and ask the more important question of why he uses so many apps in the first place. What he needs—what all of us who struggle with these issues need—is a *philosophy of technology use*, something that covers from the ground up which digital tools we allow into our life, for what reasons, and under what constraints. In the absence of this introspection, we'll be left struggling in a whirlwind of addictive and appealing cyber-trinkets, vainly hoping that the right mix of ad hoc hacks will save us.

As I mentioned in the introduction, I have one such philosophy to propose:

Digital Minimalism

A philosophy of technology use in which you focus your online time on a small number of carefully selected and optimized activities that strongly support things you value, and then happily miss out on everything else.

The so-called digital minimalists who follow this philosophy constantly perform implicit cost-benefit analyses. If a new
technology offers little more than a minor diversion or trivial
convenience, the minimalist will ignore it. Even when a new
technology promises to support something the minimalist
values, it must still pass a stricter test: Is this the *best* way to use
technology to support this value? If the answer is no, the minimalist will set to work trying to optimize the tech, or search
out a better option.

By working backward from their deep values to their technology choices, digital minimalists transform these innovations from a source of distraction into tools to support a life
well lived. By doing so, they break the spell that has made so
many people feel like they're losing control to their screens.

Notice, this minimalist philosophy contrasts starkly with
the maximalist philosophy that most people deploy by default—
a mind-set in which *any* potential for benefit is enough to start
using a technology that catches your attention. A maximalist is
very uncomfortable with the idea that anyone might miss out
on something that's the least bit interesting or valuable. Indeed, when I first started writing publicly about the fact that
I've never used Facebook, people in my professional circles
were aghast for exactly this reason. "Why do I need to use
Facebook?" I would ask. "I can't tell you exactly," they would
respond, "but what if there's something useful to you in there
that you're missing?"

This argument sounds absurd to digital minimalists, because they believe that the best digital life is formed by care-

fully curating their tools to deliver massive and unambiguous benefits. They tend to be incredibly wary of low-value activities that can clutter up their time and attention and end up hurting more than they help. Put another way: minimalists don't mind missing out on small things; what worries them much more is diminishing the large things they *already know for sure* make a good life good.

To make these abstract ideas more concrete, let's consider some real-world examples of digital minimalists I uncovered in my research on this emerging philosophy. For some of these minimalists, the requirement that a new technology strongly supports deep values led to the rejection of services and tools that our culture commonly believes to be mandatory. Tyler, for example, originally joined the standard social media services for the standard reasons: to help his career, to keep him connected, and to provide entertainment. Once Tyler embraced digital minimalism, however, he realized that although he valued all three of these goals, his compulsive use of social networks offered at best minor benefits, and did not qualify as the *best* way to use technology for these purposes. So he quit all social media to pursue more direct and effective ways to help his career, connect with other people, and be entertained.

I met Tyler roughly a year after his minimalist decision to leave social media. He was clearly excited by how his life had changed during this period. He started volunteering near his home, he exercises regularly, he's reading three to four books a month, he began to learn to play the ukulele, and he told me that now that his phone is no longer glued to his hand, he's closer than he has ever been with his wife and kids. On the professional side, the increased focus he achieved after leaving these services earned him a promotion. "Some of my work clients have noticed a change in me and they will ask what I am doing differently," he told me. "When I tell them I quit social media, their response is 'I wish I could do that, but I just can't.' The reality, however, is that they literally have no good reason to be on social media!"

As Tyler is quick to admit, he can't completely attribute all of these good things to his specific decision to quit social media. In theory, he could have still learned the ukulele or spent more time with his wife and kids while maintaining a Facebook account. His decision to leave these services, how-ever, was about more than a tweak to his digital habits; it was a symbolic gesture that reinforced his new commitment to the minimalist philosophy of working backward from your deeply held values when deciding how to live your life.

Adam provides another good example of this philosophy leading to the rejection of a technology that we've been told is fundamental. Adam runs a small business, and the ability to remain connected to his employees is important for his liveli-hood. Recently, however, he became worried about the exam-ple he was setting for his nine- and thirteen-year-old kids. He could talk to them about the importance of experiencing life beyond a glowing screen, he realized, but the message wouldn't stick until they saw him demonstrating this behavior in his own life. So he did something radical: he got rid of his smart-phone and replaced it with a basic flip phone.

"I have never had a better teachable moment in my life," he told me about his decision. "My kids know my business depends on a smart device and saw how much I used it, and here I was giving it up?! I was able to clearly explain why, *and they got it!*"

As Adam admits, the loss of his smartphone made certain things in his work life more annoying. In particular, he relies heavily on text messages to coordinate with his staff, and he soon relearned how hard it is to type on the little plastic buttons of an old-fashioned cell phone. But Adam is a digital minimalist, which means maximizing convenience is prioritized much lower than using technology to support his values. As a father, teaching his kids an important lesson about embracing life beyond the screen was far more important than faster typing.

Not all digital minimalists end up completely rejecting common tools. For many, the core question of "is this the *best* way to use technology to support this value?" leads them to carefully optimize services that most people fiddle with mindlessly.

Michal, for example, decided her obsession with online media was causing more harm than good. In response, she restricted her digital information intake to a pair of email newsletter subscriptions and a handful of blogs that she checks "less than once a week." She told me that these carefully selected feeds still satisfy her craving for stimulating ideas and information without dominating her time and toying with her mood.

Another digital minimalist named Charles told me a similar story. He had been a Twitter addict before adopting this philosophy. He has since quit that service and instead receives his news through a curated collection of online magazines that he checks once a day in the afternoon. He told me that he's better informed than he was during his Twitter days while also now thankfully freed of the addictive checking and refreshing that Twitter encourages in its users.

Digital minimalists are also adept at stripping away superfluous features of new technologies to allow them to access functions that matter while avoiding unnecessary distraction. Carina, for example, is on the executive council of a student organization that uses a Facebook group to coordinate its activities. To prevent this service from exploiting her attention every time she logs on for council business, she reduced her set of friends down to only the fourteen other people on the executive council and then unfollowed them. This preserves her ability to coordinate on the Facebook group while at the same time keeping her newsfeed empty.

Emma found a different approach to a similar end when she discovered that she could bookmark the Facebook notifications screen, allowing her to jump straight to the page that shows posts from a graduate student group she follows—bypassing the service's most distracting features. Blair did something similar: bookmarking the Facebook events page so she could check on upcoming community events while bypassing "[all the] junk that Facebook is made up of." Blair told me that keeping up with local events through this bookmarked page takes about five minutes, once or twice a week. Carina and Emma report similarly miniscule times spent using the service. The *average* Facebook user, by contrast, uses the company's products a little over fifty minutes per day. These optimizations might seem small, but they yield a major difference in these digital minimalists' daily lives.

A particularly heartwarming example of digital minimalism unlocking new value is the story of Dave, a creative director and father of three. After embracing minimalism, Dave reduced his persistent social media use down to only a single service, Instagram, which he felt offered significant benefits to his deep interest in art. In true minimalist fashion, however, Dave didn't settle for simply deciding to "use" Instagram; he instead thought hard about how *best* to integrate this tool into his life. In the end, he settled on posting one picture every week of whatever personal art project he happens to be working on. "It's a great way for me to have a visual archive of my projects," he explained. He also follows only a small number of accounts, all of which belong to artists whose work inspires him—making the experience of checking his feed both fast and meaningful.

The reason I like Dave's story, however, is what was enabled by his decision to significantly cut back on how much he uses these services. As Dave explained to me, his own father wrote him a handwritten note every week during his freshman year of college. Still touched by this gesture, Dave began a habit of drawing a new picture every night to place in his oldest daughter's lunchbox. His two youngest children watched this ritual with interest. When they became old enough for lunchboxes, they were excited to start receiving their daily drawings as well. "Fast-forward a couple of years, and I'm spending a decent chunk of time every night doing three drawings!" Dave told me with obvious pride. "This wouldn't have been possible if I didn't protect how I spend my time."

THE PRINCIPLES OF DIGITAL MINIMALISM

So far in this chapter, I've argued that the best way to fight the tyranny of the digital in your life is to embrace a philosophy of technology use based in your deeply held values. I then proposed digital minimalism as one such philosophy, and provided examples of it in action. Before I can ask you to experiment with digital minimalism in your own life, however, I must first provide you with a more thorough explanation for *why* it works. My argument for this philosophy's effectiveness rests on the following three core principles:

- **Principle #1:** *Clutter is costly.*

    Digital minimalists recognize that cluttering their time and attention with too many devices, apps, and services creates an overall negative cost that can swamp the small benefits that each individual item provides in isolation.

- **Principle #2:** *Optimization is important.*

  Digital minimalists believe that deciding a particular technology supports something they value is only the first step. To truly extract its full potential benefit, it's necessary to think carefully about *how* they'll use the technology.

- **Principle #3:** *Intentionality is satisfying.*

  Digital minimalists derive significant satisfaction from their general commitment to being more intentional about how they engage with new technologies. This source of satisfaction is independent of the specific decisions they make and is one of the biggest reasons that minimalism tends to be immensely meaningful to its practitioners.

The validity of digital minimalism is self-evident once you accept these three principles. With this in mind, the remainder of this chapter is dedicated to proving them true.

AN ARGUMENT FOR PRINCIPLE #1:
THOREAU'S NEW ECONOMICS

Near the end of March in 1845, Henry David Thoreau borrowed an ax and walked into the woods near Walden Pond. He felled young white pine trees, which he hewed into studs and rafters and floorboards. Using more borrowed tools, he notched mortise and tenon joints and assembled these pieces into the frame of a modest cabin.

Thoreau was not hurried in these efforts. Each day he brought with him a lunch of bread and butter wrapped in newspaper, and after eating his meal he would read the wrapping. He found time during this leisurely construction process to take detailed notes on the nature that surrounded him. He observed the properties of the late season ice on the pond and the fragrance of the pine pitch. One morning while soaking a hickory wedge in the cold pond water, he saw a striped snake slide into the pond and lay still on the bottom. He watched it for over a quarter of an hour.

In July, Thoreau moved into the cabin where he then lived for the next two years. In the book *Walden*, he wrote about this experience, famously describing his motivation as follows: "I went to the woods because I wished to live deliberately, to front only the essential facts of life, and see if I could not learn what it had to teach, and not, when I came to die, discover that I had not lived."

Over the ensuing decades, as Thoreau's ideas diffused through pop culture and people became less likely to confront his actual text, his experiment at Walden Pond has taken on a poetic tinge. (Indeed, the passion-seeking boarding school students in 1989's *Dead Poets Society* open their secret poetry reading meetings by reciting the "deliberate living" quote from *Walden*.) Thoreau, we imagine, was seeking to be transformed by the subjective experience of living deliberately—planning to walk out of the woods changed by transcendence. There's truth to this interpretation, but it misses a whole other side to Thoreau's experiment. He had also been working out a new theory of economics that attempted to push back against the worst dehumanizing effects of industrialization. To help validate his theory, he needed more data, and his time spent by the pond was designed in large part to become a source of this needed information. It's important for our purposes to understand this more pragmatic side to *Walden*, as Thoreau's often overlooked economic theory provides a powerful justification for our first principle of minimalism: that more can be less.

■ ■ ■

The first and longest chapter of *Walden* is titled "Economy." It contains many of Thoreau's signature poetic flourishes about nature and the human condition. It also, however, contains a surprising number of bland expense tables, recording costs down to a fraction of a cent, such as the following:

House	$28.12 ½
Farm one year	14.72 ½
Food eight months	8.74
Clothing, etc., eight months	8.40 ¾
Oil, etc., eight months	2.00
In all	$61.99 ¾

Thoreau's purpose in these tables is to capture precisely (not poetically or philosophically) how much it cost to support his life at Walden Pond—a lifestyle that, as he argues at length in this first chapter, satisfies all the basic human needs: food, shelter, warmth, and so on. Thoreau then contrasts these costs with the hourly wages he could earn with his labor to arrive at the final value he cared most about: How much of his *time* must be sacrificed to support his minimalist lifestyle? After plugging in the numbers gathered during his experiment, he determined that hiring out his labor only one day per week would be sufficient.

This magician's trick of shifting the units of measure from money to time is the core novelty of what the philosopher Frédéric Gros calls Thoreau's "new economics," a theory that builds on the following axiom, which Thoreau establishes early in *Walden*: "The cost of a thing is the amount of what I will call life which is required to be exchanged for it, immediately or in the long run."

This new economics offers a radical rethinking of the consumerist culture that began to emerge in Thoreau's time. Standard economic theory focuses on monetary outcomes. If working one acre of land as a farmer earns you $1 a year in profit, and working sixty acres earns you $60, then you should, if it's at all possible, work the sixty acres—it produces strictly more money.

Thoreau's new economics considers such math woefully incomplete, as it leaves out the cost in *life* required to achieve that extra $59 in monetary profit. As he notes in *Walden*, working a large farm, as many of his Concord neighbors did, required large, stressful mortgages, the need to maintain numerous pieces of equipment, and endless, demanding labor. He describes these farmer neighbors as "crushed and smothered under [their] load" and famously lumps them into the "mass of men lead[ing] lives of quiet desperation."

Thoreau then asks what benefits these worn-down farmers receive from the extra profit they eke out. As he proved in his Walden experiment, this extra work is not enabling the farmers to escape savage conditions: Thoreau was able to satisfy all of his basic needs quite comfortably with the equivalent of one day of work per week. What these farmers are actually gaining from all the life they sacrifice is slightly nicer stuff: venetian blinds, a better quality copper pot, perhaps a fancy wagon for traveling back and forth to town more efficiently.

When analyzed through Thoreau's new economics, this exchange can come across as ill conceived. Who could justify trading a lifetime of stress and backbreaking labor for better blinds? Is a nicer-looking window treatment really worth so much of your life? Similarly, why would you add hours of extra labor in the fields to obtain a wagon? It's true that it takes more time to walk to town than to ride in a wagon, Thoreau notes, but these walks still likely require less time than the extra work hours needed to afford the wagon. It's exactly these types of calculations that lead Thoreau to observe sardonically: "I see young men, my townsmen, whose misfortune it is to have inherited farms, house, barns, cattle, and farming tools; for these are more easily acquired than got rid of."

Thoreau's new economics was developed in an industrial age, but his basic insights apply just as well to our current digital context. The first principle of digital minimalism presented earlier in this chapter states that clutter is costly. Thoreau's new economics helps explain why.

When people consider specific tools or behaviors in their digital lives, they tend to focus only on the value each produces. Maintaining an active presence on Twitter, for example, might occasionally open up an interesting new connection or expose you to an idea you hadn't heard before. Standard economic thinking says that such profits are good, and the more you receive the better. It therefore makes sense to clutter your digital life with as many of these small sources of value as you can find, much as it made sense for the Concord farmer to cultivate as many acres of land as he could afford to mortgage.

Thoreau's new economics, however, demands that you balance this profit against the costs measured in terms of "your life." How much of your time and attention, he would ask, must be sacrificed to earn the small profit of occasional connections and new ideas that is earned by cultivating a significant presence on Twitter? Assume, for example, that your Twitter habit effectively consumes ten hours per week. Thoreau would note that this cost is almost certainly way too high for the limited benefits it returns. If you value new connections and exposure to interesting ideas, he might argue, why not adopt a habit of attending an interesting talk or event every month, and forcing yourself to chat with at least three people while there? This would produce similar types of value but consume only a few hours of your life per month, leaving you with an extra thirty-seven hours to dedicate to other meaningful pursuits.

These costs, of course, also tend to compound. When you combine an active Twitter presence with a dozen other attention-demanding online behaviors, the cost in life becomes extreme. Like Thoreau's farmers, you end up "crushed and smothered" under the demands on your time and attention, and in the end, all you receive in return for sacrificing so much of your life is a few nicer trinkets—the digital equivalent of the farmer's venetian blinds or fancier pot—many of which, as shown in the Twitter example above, could probably be approximated at a much lower cost, or eliminated without any major negative impact.

This is why clutter is dangerous. It's easy to be seduced by the small amounts of profit offered by the latest app or service, but then forget its cost in terms of the most important resource we possess: the minutes of our life. This is also what makes Thoreau's new economics so relevant to our current moment. As Frédéric Gros argues:

> The striking thing with Thoreau is not the actual content of the argument. After all, sages in earliest Antiquity had already proclaimed their contempt for

possessions. . . . What impresses is the form of the
argument. For Thoreau's obsession with calculation
runs deep. . . . He says: keep calculating, keep weigh-
ing. What exactly do I gain, or lose?

Thoreau's obsession with calculation helps us move past the
vague subjective sense that there are trade-offs inherent in digital clutter, and forces us instead to confront it more precisely.
He asks us to treat the minutes of our life as a concrete and
valuable substance—arguably the most valuable substance we
possess—and to always reckon with how much of this life we
trade for the various activities we allow to claim our time.
When we confront our habits through this perspective, we will
reach the same conclusion now that Thoreau did in his era:
more often than not, the cumulative cost of the noncrucial
things we clutter our lives with can far outweigh the small benefits each individual piece of clutter promises.

AN ARGUMENT FOR PRINCIPLE #2:
THE RETURN CURVE

The law of diminishing returns is familiar to anyone who
studies economics. It applies to the improvement of production
processes and says, at a high level, that investing more resources into a process cannot indefinitely improve its output—
eventually you'll approach a natural limit and start experiencing
less and less extra benefit from continued investment.

A classic example from economics textbooks concerns workers on a hypothetical automobile assembly line. At first, as you increase the number of workers, you generate large increases in the rate at which finished cars come off the line. If you continue to assign more workers to the line, however, these improvements will get smaller. This might happen for many reasons. Perhaps, for example, you begin to run out of space to add the new workers, or other limiting factors, like the maximum speed of the conveyer belt, come into play.

If you plot this law for a given process and resource, with value produced on the y-axis and amount of resource invested on the x-axis, you'll encounter a familiar curve. At first, as additional increases in resources cause rapid improvements in output, the curve rises quickly, but over time, as the returns diminish, the curve flattens out. The exact parameters of this *return curve* vary between different processes and resources, but its general shape is shared by many scenarios—a reality that has made this law a fundamental component of modern economic theory.

The reason I'm introducing this idea from economics in this chapter on digital minimalism is the following: if you're willing to accept some flexibility in your definition of "production process," the law of diminishing returns can apply to the various ways in which we use new technologies to produce value in our personal lives. Once we view these *personal technology processes* through the perspective of diminishing returns, we'll gain the precise vocabulary we need to understand the validity of the second principle of minimalism, which states that optimizing how we use technology is just as important as how we choose what technologies to use in the first place.

■ ■ ■

When considering personal technology processes, let's focus in particular on the energy invested in trying to improve the value these processes return in your life, for example, through better selection of tools or the adoption of smarter strategies for using the tools. If you increase the amount of energy you invest into this optimization, you'll increase the amount of value the process returns. At first, these increases will be large. As the law of diminishing returns tells us, however, eventually these increases will diminish as you approach a natural limit.

To make this more concrete, let's work through a brief hypothetical example. Assume that you find it important to remain informed about current events. New technologies can certainly help you support this goal. Perhaps, at first, the process you deploy is just keeping an eye on the links that pop up in your social media feeds. This process produces some value, as it keeps you more informed than if you weren't using the internet at all for this purpose, but it leaves a lot of room for improvement.

With this in mind, assume you invest some energy to identify a more carefully curated set of online news sites to follow, and to find an app, like Instapaper, that allows you to clip articles from these sites and read them all together in a nice interface that culls distracting ads. This improved personal technology process for keeping informed is now producing even more value in your personal life. Perhaps, as the final step in this optimization, you discover through trial and error that you're best able to absorb complex articles when you clip them throughout the week and then sit down to read through them all on Saturday morning on a tablet over coffee at a local café.

At this point, your optimization efforts have massively increased the value you receive from this personal technology process for staying informed. You can now stay up to date in a pleasing manner that has a limited impact on your time and attention during the week. As the law of diminishing returns tells us, however, you're probably nearing the natural limit, after which improving this process further will become increasingly difficult. Put more technically: you've reached the later part of the return curve.

The reason the second principle of minimalism is so important is that most people invest very little energy into these types of optimizations. To use the appropriate economic terminology, most people's personal technology processes currently exist on the *early* part of the return curve—the location where additional attempts to optimize will yield massive improvements. It's this reality that leads digital minimalists to embrace the second principle, and focus not just on *what* technologies they adopt, but also on *how* they use them.

The example I gave above was hypothetical, but you find similar instances of optimization producing big returns when you study the stories of real-world digital minimalists. Gabriella, for example, signed up for Netflix as a better (and cheaper) source of entertainment than cable. She became prone, how-

ever, to binge-watching, which hurt her professional productivity and left her feeling unfulfilled. After some further
experimentation, Gabriella adopted an optimization to this
process: she's not allowed to watch Netflix alone.* This restriction still allows her to enjoy the value Netflix offers, but
to do so in a more controlled manner that limits its potential
for abuse and strengthens something else she values: her social
life. "Now [streaming shows is] a social activity instead of an
isolating activity," she told me.

Another optimization that was common among the digital
minimalists I studied was to remove social media apps from
their phones. Because they can still access these sites through
their computer browsers, they don't lose any of the high-value
benefits that keep them signed up for these services. By removing the apps from their phones, however, they eliminated
their ability to browse their accounts as a knee-jerk response
to boredom. The result is that these minimalists dramatically
reduced the amount of time they spend engaging with these
services each week, while barely diminishing the value they
provide to their lives—a much better personal technology
process than thoughtlessly tapping and swiping these apps
throughout the day as the whim strikes.

There are two major reasons why so few people have both-
ered to adopt the bias toward optimization exhibited by

---

* Gabriella is not alone in this optimization. I was surprised to discover
multiple digital minimalists (usually young people) who found a good balance by restricting streaming entertainment to social situations.

Gabriella or the minimalists who streamlined their social media experience. The first is that most of these technologies are still relatively new. Because of this reality, their role in your life can still seem novel and fun, obscuring more serious questions about the specific value they're providing. This freshness, of course, is starting to fade as the smartphone and social media era advances beyond its heady early years, which will lead people to become increasingly impatient with the shortcomings of their unpolished processes. As the author Max Brooks quipped in a 2017 TV appearance, "We need to reevaluate [our current relationship with] online information sort of the way we reevaluated free love in the 80s."

The second reason so few think about optimizing their technology use is more cynical: The large attention economy conglomerates that introduced many of these new technologies don't want us thinking about optimization. These corporations make more money the more time you spend engaged with their products. They want you, therefore, to think of their offerings as a sort of fun ecosystem where you mess around and interesting things happen. This mind-set of general use makes it easier for them to exploit your psychological vulnerabilities.

By contrast, if you think of these services as offering a collection of features that you can carefully put to use to serve specific values, then almost certainly you'll spend much less time using them. This is why social media companies are purposely vague in describing their products. The Facebook mission statement, for example, describes their goal as "giv[ing] people the power to build community and bring the world closer together." This goal is generically positive, but how exactly you use Facebook to accomplish it is left underspecified. They hint that you just need to plug into their ecosystem and start sharing and connecting, and eventually good things will happen.

Once you break free from this mind-set, however, and begin seeing new technologies simply as tools that you can deploy selectively, you're able to fully embrace the second principle of minimalism and start furiously optimizing— enabling you to reap the advantages of vaulting up the return curve. Finding useful new technologies is just the first step to improving your life. The real benefits come once you start experimenting with how best to use them.

AN ARGUMENT FOR PRINCIPLE #3: THE LESSONS OF THE AMISH HACKER

The Amish complicate any serious discussion of modern technology's impact on our culture. The popular understanding of this group is that they're frozen in time—reluctant to adopt any tools introduced after the mid-eighteenth-century period when they first began settling in America. From this perspective, these communities are mainly interesting as a living museum of an older age, a quaint curiosity.

But then you start talking to scholars and writers who study the Amish seriously, and you begin to hear confusing state-ments that muddy these waters. John Hostetler, for exam-ple, who literally wrote the book on their society, claims the following: "Amish communities are not relics of a bygone era. Rather, they are demonstrations of a different form of moder-nity." The technologist Kevin Kelly, who spent a significant amount of time among the Lancaster County Amish, goes even further, writing: "Amish lives are anything but antitech-nological. In fact, on my several visits with them, I have found them to be ingenious hackers and tinkers, the ultimate makers and do-it-yourselvers. They are often, surprisingly, pro-technology."

As Kelly elaborates in his 2010 book, *What Technology Wants*, the simple notion of the Amish as Luddites vanishes as soon as you approach a standard Amish farm, where "cruis-ing down the road you may see an Amish kid in a straw hat and suspenders zipping by on Rollerblades." Some Amish communities use tractors, but only with metal wheels so they cannot drive on roads like cars. Some allow a gas-powered wheat thresher but require horses to pull the "smoking, noisy contraption." Personal phones (cellular or household) are al-most always prohibited, but many communities maintain a community phone booth.

Almost no Amish communities allow automobile owner-ship, but it's typical for Amish to travel in cars driven by oth-ers. Kelly reports that the use of electricity is common, but it's usually forbidden to connect to the larger municipal power grid. Disposable diapers are popular, as are chemical fertilizers. In one memorable passage, Kelly talks about visiting a family that uses a $400,000 computer-controlled precision milling machine to produce pneumatic parts needed by the community. The machine is run by the family's bonnet-wearing, ten-year-old daughter. It's housed behind their horse stable.

Kelly, of course, is not the only person to notice the Amish's complicated relationship with modern technologies. Donald Kraybill, a professor at Elizabethtown College who co-authored a book on the Amish, emphasizes the changes that have occurred as more members of these communities embrace entrepreneurship over farming. He talks about an Amish woodshop with nineteen employees who use drills, saws, and nail guns, but instead of receiving power from the electric grid, they use solar panels and diesel generators. Another Amish entrepreneur has a website for his business, but it's maintained by an outside firm. Kraybill has a term for the nuanced and sometimes contrived ways that these start-ups use technology: "Amish hacking."

These observations dismiss the popular belief that the Amish reject all new technologies. So what's really going on here? The Amish, it turns out, do something that's both shockingly radical and simple in our age of impulsive and complicated consumerism: they start with the things they value most, then work backward to ask whether a given new technology performs more harm than good with respect to these values. As Kraybill elaborates, they confront the following questions: "Is this going to be helpful or is it going to be detrimental? Is it going to bolster our life together, as a community, or is it going to somehow tear it down?"

When a new technology rolls around, there's typically an "alpha geek" (to use Kelly's term) in any given Amish community that will ask the parish bishop permission to try it out. Usually the bishop will agree. The whole community will then observe this first adopter "intently," trying to discern the ultimate impact of the technology on the things the community values most. If this impact is deemed more negative than helpful, the technology is prohibited. Otherwise it's allowed, but usually with caveats on its use that optimize its positives and minimize its negatives.

The reason most Amish are prohibited from owning cars, for example, but are allowed to drive in motor vehicles driven by other people, has to do with the impact of owning an automobile on the social fabric of the community. As Kelly explains: "When cars first appeared at the turn of the last century, the Amish noticed that drivers would leave the community to go picnicking or sightseeing in other towns, instead of visiting family or the sick on Sundays, or patronizing local shops on Saturday." As a member of an Amish community explained to Kraybill during his research: "When people leave the Amish, the first thing they do is buy a car." So owning a car is banned in most parishes.

This type of thinking also explains why an Amish farmer can own a solar panel or run power tools on a generator but cannot connect to the power grid. The problem is not electricity; it's the fact that the grid connects them too strongly to the world outside of their local community, violating the Amish commitment to the biblical tenet to "be in the world, but not of it."

Once you encounter this more nuanced approach to technology, you can no longer dismiss the Amish lifestyle as a quaint curiosity. As John Hostetler explained, their philosophy is not a rejection of modernity, but a "different form" of it. Kevin Kelly goes a step further and claims that it's a form of modernity that we cannot ignore given our current struggles. "In any discussion about the merits of avoiding the addictive grasp of technology," he writes, "the Amish stand out as offering an honorable alternative." It's important to understand what exactly makes this alternative honorable, as it's in these advantages that we'll uncover a strong argument for the third principle of minimalism, which claims that approaching decisions with intention can be more important than the impact of the actual decisions themselves.

■ ■ ■

At the core of the Amish philosophy regarding technology is the following trade-off: The Amish prioritize the benefits generated by acting intentionally about technology over the benefits lost from the technologies they decide not to use. Their gamble is that *intention trumps convenience*—and this is a bet that seems to be paying off. The Amish have remained a relatively stable presence in America for over two hundred years of rapid modernity and cultural upheavals. Unlike some religious sects that attempt to entrap members through threats and denial of connection to the outside world, the Amish still practice *Rumspringa*. During this ritual period, which begins at the age of sixteen, Amish youth are allowed to leave home and experience the outside world beyond the restrictions of their community. It is then their decision, after having seen what they will be giving up, whether or not they accept baptism into the Amish church. By one sociologist's calculations, the percentage of Amish youth that decide to stay after Rumspringa is in the range of 80 to 90 percent.

We should be careful, however, not to push the Amish example too far as a case study for meaningful living. The restrictions that guide each community, called the *Ordnung*, are typically decided and enforced by a group of four men—a bishop, two ministers, and a deacon—who serve for life. There's a communion ceremony performed twice a year in which complaints about the Ordnung can be aired and consensus pursued, but many in these communities, including, notably, women, can remain largely disenfranchised.

From this perspective, the Amish underscore the principle that acting intentionally with respect to technology can be a standalone source of value, but their example leaves open the question of whether this value persists even when we eliminate the more authoritarian impulses of these communities. Fortunately, we have good reasons to believe it does.

A useful thought experiment along these lines is to consider the closely related Mennonite Church. Like the Amish, Mennonites embrace the biblical principle to *be in the world, but not of it*, which leads to a similar embrace of simplicity and a suspicion of cultural trends that threaten core values of maintaining strong communities and virtuous living. Unlike the Amish, however, the Mennonites include more liberal members who integrate with the broader society, taking on *personal* responsibility for making decisions in a way that's consistent with their church's principles. This creates an opportunity to see Amish-style values toward technology applied in the absence of an authoritarian Ordnung.

Curious to encounter this philosophy in action, I set up a conversation with a liberal Mennonite named Laura, a schoolteacher who lives with her husband and daughter in Albuquerque, New Mexico. Laura attends a local Mennonite church and lives in a neighborhood with at least a dozen other Mennonite families, which keeps her connected to this community's values. But decisions about her lifestyle are hers alone. This latter point hasn't stopped her from acting with intention regarding her technology choices. This reality is best emphasized by what is arguably her most radical decision: she has never owned a smartphone and has no intention of buying one.

"I don't think I'd be a good smartphone user," she explained to me. "I don't trust myself to just let it be and not think about it. When I leave the house, I don't think about all of these distractions. I'm free from it." Most people, of course, would dismiss the possibility of ditching their phone by listing all the different things its makes (slightly) easier—from looking up a restaurant review in a new city to using GPS directions. The loss of these small dollops of value doesn't concern Laura. "Writing down directions before leaving home is not a big deal for me," she said. What Laura does care about is the way her intentional decision supports things she finds massively valuable, such as her ability to connect with people she cares about and enjoy life in the moment. In our conversation, she emphasized the importance of being present with her daughter, even when bored, and the value she gets out of spending time with friends free from distraction. Laura also connects efforts to be a "conscientious consumer" with issues relating to social justice, which also play a big role in the Mennonite Church.

As with the Amish who find contentment without modern conveniences, an important source of Laura's satisfaction with her smartphone-free life comes from the choice itself. "My decision [to not use a smartphone] gives me a sense of autonomy," she told me. "I'm controlling the role technology is allowed to play in my life." After a moment of hesitation, she adds: "It makes me feel a little smug at times." What Laura describes modestly as smugness is almost certainly something more fundamental to human flourishing: the sense of meaning that comes from acting with intention.

■ ■ ■

Pulling together these pieces, we arrive at a strong justification for the third principle of minimalism. Part of what makes this philosophy so effective is that the very act of being selective about your tools will bring you satisfaction, typically much more than what is lost from the tools you decide to avoid.

I tackled this principle last because its lesson is arguably the most important. As demonstrated by the Old Order Amish farmer happily riding a horse-drawn buggy, or the urban Mennonite content with her old-fashioned cell phone, it's the commitment to minimalism itself that yields the bulk of their satisfaction. The sugar high of convenience is fleeting and the sting of missing out dulls rapidly, but the meaningful glow that comes from taking charge of what claims your time and attention is something that persists.

A NEW LOOK AT OLD ADVICE

The central idea of minimalism, that less can be more, is not novel. As mentioned in the introduction, this concept dates back to antiquity and has been repeatedly espoused since. The fact, therefore, that this old idea might apply to the new technologies that define so much about our current age shouldn't be surprising.

That being said, the past couple of decades are also defined by a resurgent narrative of techno-maximalism that contends more is better when it comes to technology—more connections, more information, more options. This philosophy cleverly dovetails with the general objective of the liberal humanism project to offer individuals more freedom, making it seem vaguely illiberal to avoid a popular social media platform or decline to follow the latest online chatter.

This connection, of course, is specious. Outsourcing your autonomy to an attention economy conglomerate—as you do when you mindlessly sign up for whatever new hot service emerges from the Silicon Valley venture capitalist class—is the opposite of freedom, and will likely degrade your individuality. But given the current strength of the maximalism argument, I felt it necessary to provide the full-throated defense of minimalism detailed in this chapter. Even old ideas require new investigation to underscore their continued relevance.

When it comes to new technologies, less almost certainly is more. Hopefully the preceding pages made it clear why this is true.

3

The Digital Declutter

ON (RAPIDLY) BECOMING MINIMALIST

Assuming I've convinced you that digital minimalism is worthwhile, the next step is to discuss how best to adopt this lifestyle. In my experience, gradually changing your habits one at a time doesn't work well—the engineered attraction of the attention economy, combined with the friction of convenience, will diminish your inertia until you backslide toward where you started.

I recommend instead a rapid transformation—something that occurs in a short period of time and is executed with enough conviction that the results are likely to stick. I call the particular rapid process I have in mind the *digital declutter*. It works as follows.

The Digital Declutter Process

1. Put aside a thirty-day period during which you will take a break from optional technologies in your life.
2. During this thirty-day break, explore and rediscover activities and behaviors that you find satisfying and meaningful.
3. At the end of the break, reintroduce optional technologies into your life, starting from a blank slate. For each technology you reintroduce, determine what value it serves in your life and how specifically you will use it so as to maximize this value.

Much like decluttering your house, this lifestyle experiment provides a reset for your digital life by clearing away distracting tools and compulsive habits that may have accumulated haphazardly over time and replacing them with a much more intentional set of behaviors, optimized, in proper minimalist fashion, to support your values instead of subverting them.

As noted earlier, the second part of this book will provide ideas and strategies for shaping your digital minimalist lifestyle into something sustainable over the long term. My suggestion, however, is to start with this declutter, and then once your transformation has begun, turn to the chapters that follow to optimize your setup. As is often true in life, getting started is the most important step. With this in mind, we'll continue by looking closer at the details of executing the digital declutter. Fortunately, as I'll explain next, when it comes to examining how best to succeed with this process, we don't have to start from scratch. Many others have trod this path before.

■ ■ ■

In early December 2017, I sent an email to my mailing list that summarized the main ideas of this process. "I'm looking for volunteers," I wrote, "who are willing to attempt a digital declutter during the month of January and provide me updates along the way." I expected forty to fifty brave readers to volunteer. My guess was wrong: over 1,600 signed up. Our efforts even made national news.

In February, I began to gather more-detailed reports from participants. I wanted to find out what rules they put in place regarding their technology use during the declutter and how they fared during the thirty-day period. I was particularly interested to hear about the decisions they made when reintroducing these technologies back into their lives.

After receiving and reviewing hundreds of these in-depth dissections, two conclusions became clear. First, the digital declutter works. People were surprised to learn the degree to which their digital lives had become cluttered with reflexive behaviors and compulsive tics. The simple action of sweeping away this detritus and starting from scratch in crafting their digital life felt like lifting a psychological weight they didn't realize had been dragging them down. They came out of the declutter with a streamlined digital lifestyle that felt, in some ineffable sense, "right."

The second clear conclusion I reached is that the declutter process is tricky. A nontrivial number of people ended up aborting this process before the full thirty days were done. Interestingly, most of these early exits had little to do with insufficient willpower—this was an audience who was self-selected based on their drive to improve. More common were subtle mistakes in implementation. A typical culprit, for example, was technology restriction rules that were either too vague or too strict. Another mistake was not planning what to *replace* these technologies with during the declutter period— leading to anxiety and boredom. Those who treated this experiment purely as a *detox*, where the goal was to simply take a break from their digital life before returning to business as usual, also struggled. A temporary detox is a much weaker resolution than trying to permanently change your life, and therefore much easier for your mind to subvert when the going gets tough.

Given the reality of this second conclusion, I will dedicate the remainder of this chapter to providing clarifying explanations and suggestions for the three steps of the declutter process summarized above. For each of these steps, I'll provide detailed examples from participants in my mass digital declutter experiment to help you avoid common pitfalls and tweak your experience to maximize your probability of success.

STEP #1: DEFINE YOUR TECHNOLOGY RULES

During the thirty days of your digital declutter, you're supposed to take a break from "optional technologies" in your life. The first step of the declutter process, therefore, is to define which technologies fall into this "optional" category.

When I say *technology* in this context, I mean the general class of things we've been calling "new technologies" throughout this book, which include apps, websites, and related digital tools that are delivered through a computer screen or a mobile phone and are meant to either entertain, inform, or connect you. Text messaging, Instagram, and Reddit are examples of the types of technologies you need to evaluate when preparing for your digital declutter; your microwave, radio, or electric toothbrush are not.

An interesting special case brought to my attention by many participants during the mass declutter experiment is video games. These can't be neatly classified under the "new technology" label as they've been around for decades before the digital network and mobile computing revolutions of the past twenty years. But many people—especially young men— feel an addictive pull to these games that's similar to what they experience from other new technologies. As a twenty-nine-year-old business owner named Joseph told me, he feels "restless without video games to occupy my downtime." He ended

up classifying these games alongside his compulsive blog consumption as factors in his digital life that were wearing him
down. If, like Joseph, you think these games are a nontrivial
part of your life, feel free to put them on the list of technologies you're evaluating when figuring out your declutter
rules.

Another borderline case is television—which, in an age of
streaming, is a vague term that can cover many different visual
entertainments. Prior to the mass declutter experiment, I was
somewhat ambivalent as to whether streaming Netflix, and its
equivalents, was something to consider as a potentially optional technology. The feedback I received from participants,
however, was near unequivocal: *You should.* As a management
consultant named Kate told me: "I have so many ideas I'd like
to implement, but every time I [sat] down to work on them,
somehow Netflix [appeared] on my screen." These technologies, participants like Kate insisted, should be included when
defining your personal declutter rules.

Once you've identified the class of technologies that are relevant, you must then decide which of them are sufficiently
"optional" that you can take a break from them for the full
thirty days of the declutter process. My general heuristic is the
following: consider the technology optional unless its temporary removal would harm or significantly disrupt the daily operation of your professional or personal life.

This standard exempts most professional technologies from
being deemed optional. If you stop checking your work
email, for example, this would harm your career—so you can't

use me as an excuse to shut down your inbox for a month. Similarly, if your job requires you to occasionally monitor Facebook Messenger to help recruit students (as was the case for a music professor named Brian who participated in my experiment), then, of course, this activity is not optional either.

On the personal side, these exemptions usually apply to technologies that play a key logistical role. If your daughter uses text messaging to tell you when she's ready to be picked up from soccer practice, then it's okay to still use text messages for this purpose. Similar exemptions also apply when a technology's removal might cause serious harm to relationships: for example, using FaceTime to talk with a spouse deployed overseas with the military.

Don't, however, confuse "convenient" with "critical." It's inconvenient to lose access to a Facebook group that announces campus events, but in a thirty-day period, this lack of information won't cause any critical damage to your social life, and it might expose you to interesting alternative uses for your time. Similarly, several participants in the mass declutter experiment claimed they needed to keep using instant message tools like WhatsApp or Facebook Messenger because it was the easiest way to keep up with friends overseas. This might be true, but in many cases, these relationships can withstand one month of less frequent contact.

More importantly, the inconvenience might prove useful. Losing lightweight contact with your international friends might help clarify which of these friendships were real in the first place, and strengthen your relationships with those who remain. This is exactly what happened with Anya, a participant in my experiment who is from Belarus but is currently studying at an American university. As she told the *New York Times* in an article about my experiment, taking a break from online socializing with her international friends helped her "feel more invested in the time I spend with people. . . . Because we [interacted] less frequently, we [had] this idea that we want to make the most of the experience." A college sophomore named Kushboo put it even simpler when he told me: "In a nutshell, I only lost touch with people I didn't need (or, in some cases, didn't even want) to be constantly in touch with."

My final suggestion is to use *operating procedures* when confronting a technology that's largely optional, with the exception of a few critical use cases. These procedures specify exactly *how* and *when* you use a particular technology, allowing you to maintain some critical uses without having to default to unrestricted access. I saw many examples of these operating procedures deployed by the participants in my mass declutter experiment.

A freelance writer named Mary, for example, wanted to take a break from constantly tending to text messages on her phone. ("I'm from a very large and very 'text-y' family," she told me.) The problem was that when her husband traveled, which he did frequently, he sometimes sent Mary messages that needed fast responses. Her solution was to reconfigure her phone to send a special alert for texts from her husband, but suppress notifications for all other messages. Similarly, an environmental consultant named Mike needed to keep up with personal emails but wanted to avoid compulsive checking, so he made the rule that he could only sign into his account from his desktop PC and not his phone.

A computer scientist named Caleb decided he could still listen to podcasts, but only on his two-hour-long daily commute. ("The thought of only listening to whatever the radio sent my way was too daunting for me," he explained.) Brooke, a self-described writer, educator, and full-time mother, decided she wanted to swear off access to the internet altogether but, to make this sustainable, added two exceptions for when she could still launch a web browser: email and buying household items on Amazon.

I also noticed a lot of creativity surrounding how people throttled back streaming media in contexts where they didn't want to eliminate it altogether. A college freshman named Ramel abstained from streaming media *except* when doing so with other people, explaining: "I did not want to isolate myself in social situations where entertainment was playing." A professor named Nathaniel, on the other hand, didn't mind high-quality entertainment in his life but worried about binge-watching, so he adopted a clever restriction: "no more than two episodes of any series per week."

I would estimate around 30 percent of the rules described by participants were caveated with operating procedures, while the remaining 70 percent were blanket bans on using

a particular technology. Generally, too many operating procedures might make the declutter experiment unwieldy, but
most people required at least a few of these more nuanced
constraints.

■ ■ ■

To summarize the main points about this step:

- The digital declutter focuses primarily on new technologies, which describes apps, sites, and tools delivered
  through a computer or mobile phone screen. You should
  probably also include video games and streaming video
  in this category.
- Take a thirty-day break from any of these technologies
  that you deem "optional"—meaning that you can step
  away from them without creating harm or major problems in either your professional or personal life. In some
  cases, you'll abstain from using the optional technology
  altogether, while in other cases you might specify a set of
  operating procedures that dictate exactly when and how
  you use the technology during the process.
- In the end, you're left with a list of banned technologies
  along with relevant operating procedures. Write this
  down and put it somewhere where you'll see it every day.
  Clarity in what you're allowed and not allowed to do
  during the declutter will prove key to its success.

STEP #2: TAKE A
THIRTY-DAY BREAK

Now that you have defined your technology rules, the next step of the digital declutter is to follow these rules for thirty days.* You'll likely find life without optional technologies challenging at first. Your mind has developed certain expectations about distractions and entertainment, and these expectations will be disrupted when you remove optional technologies from your daily experience. This disruption can feel unpleasant.

Many of the participants in my mass declutter experiment, however, reported that these feelings of discomfort faded after a week or two. Brooke described this experience as follows:

> The first few days were surprisingly hard. My addictive habits were revealed in striking clarity. Moments of waiting in line, moments between activities, moments of boredom, moments I ached to check in on my favorite people, moments I wanted an escape, moments I just wanted to "look something up," moments I just needed some diversion: I'd reach for my phone and then remember that everything was gone.

---

* Obviously, your declutter does not have to span *exactly* thirty days. It's often convenient, for example, to connect the experiment to a calendar month, which means you might use thirty-one days, or perhaps twenty-eight days, depending on the month in which you run the process.

But then things got better. "As time wore on, the detox symptoms wore off and I began to forget about my phone," she explained.

A young management consultant named Daria admitted that during the first days of the experiment she would compulsively pull out her phone before realizing she had removed all of the social media and news apps. The only thing left on her phone that she could check for new information was the weather. "In that first week," she told me, "I knew the hourly weather conditions in three to four different cities"—the compulsion to browse *something* was too strong to ignore. After two weeks, however, she reported: "I lost almost any interest [in checking things online]."

This detox experience is important because it will help you make smarter decisions at the end of the declutter when you reintroduce some of these optional technologies to your life. A major reason that I recommend taking an extended break before trying to transform your digital life is that without the clarity provided by detox, the addictive pull of the technologies will bias your decisions. If you decide to reform your relationship with Instagram right this moment, your decisions about what role it should play in your life will likely be much weaker than if you instead spend thirty days without the service before making these choices.

As I mentioned earlier in this chapter, however, it's a mistake to think of the digital declutter as *only* a detox experience. The goal is not to simply give yourself a break from technology, but to instead spark a permanent transformation of your digital life. The detoxing is merely a step that supports this transformation.

With this in mind, you have duties during the declutter beyond following your technology rules. For this process to succeed, you must also spend this period trying to rediscover what's important to you and what you enjoy outside the world of the always-on, shiny digital. Figuring this out *before* you begin reintroducing technology at the end of this declutter process is crucial. An argument I'll elaborate on in part 2 of this book is that you're more likely to succeed in reducing the role of digital tools in your life if you cultivate high-quality alternatives to the easy distraction they provide. For many people, their compulsive phone use papers over a void created by a lack of a well-developed leisure life. Reducing the easy distraction without also filling the void can make life unpleasantly stale—an outcome likely to undermine any transition to minimalism.

Another reason it's important to spend the thirty days of the declutter rediscovering what you enjoy is that this information will guide you during the reintroduction of technology at the end of the process. As stated, the goal of the reintroduction is to put technology to work on behalf of specific things you value. This *means to an end* approach to technology requires clarity on what these ends actually are.

The good news is that the participants in my mass declutter experiment found it easier than expected to reconnect to the types of activities they used to enjoy before they were subverted by their screens. A graduate student named Unaiza was spending her evenings browsing Reddit. During her declutter, she redirected this time toward reading books that she borrowed from both her school and local library. "I finished eight and a half books that month," she told me. "I could never have thought about doing that before." An insurance agent named Melissa finished "only" three books during her thirty days, but also organized her wardrobe, set up dinners with friends, and scheduled more face-to-face conversations with her brother. "I wish he was participating in a declutter experiment too," she told me, "because he annoyingly watched his phone the whole time we were talking." She even kicked off a hunt for a new home that she had been delaying due to a perceived lack of time. By the end of the declutter, she had made an offer on a house, which was accepted.

Kushboo finished five books during his declutter. This was a big deal for him, as these were the first books he had read voluntarily in over three years. He also restarted his once cherished painting and computer coding hobbies. "I loved these activities," he explained, "but I stopped doing them once I started school because I thought I didn't have enough time." Caleb's search for intentional analog activities led him to start journaling and reading before bed every night. He also started listening to records on a record player, from beginning to end, with no earbuds in his ears or skip buttons to tap when antsy—which turns out to be a much richer experience than Caleb's normal habit of firing up Spotify and seeking out the perfect track. A full-time mom named Marianna became so engaged in creative pursuits during her declutter that she decided she would start her own blog to share her work and connect with other artists. An engineer named Craig reported: "Last week I actually visited my local library again for the first time since my kids have grown. . . . I was delighted to discover seven different books that seemed interesting."

Like several other parents who participated in my experiment, Tarald invested his newfound time and attention in his family. He was unhappy with how distracted he was when spending time with his sons. He told me about how, on the playground, when they would come seeking recognition for something they figured out and were proud of, he wouldn't notice, as his attention was on his phone. "I started thinking about how many of these small victories I miss out on because I feel this ridiculous need to check the news for the umpteenth time," he told me. During his declutter he rediscovered the satisfaction of spending real time with his boys instead of just spending time near them with his eyes on the screen. He noted how surreal it can feel to be the only parent at the playground who is not looking down.

Brooke also found herself "interacting more intentionally" with her kids. For her, this change wasn't engineered, but was instead a natural side effect of the declutter, which made her life feel "far less rushed and distracted"—leaving room to gravitate toward more important pursuits. She also ended up playing the piano again and relearning how to sew—underscoring the sheer quantity of the time that can be reclaimed when you sidestep mindless digital activity to once again prioritize the real you.

Brooke captures well the experience many reported about their monthlong declutter when she told me: "Stepping away for thirty-one days provided clarity I didn't know I was missing. . . . As I stand here now from the outside looking in, I see there is so much more the world has to offer!"

■ ■ ■

To summarize the main points about this step:

- You will probably find the first week or two of your digital declutter to be difficult, and fight urges to check technologies you're not allowed to check. These feelings, however, will pass, and this resulting sense of detox will prove useful when it comes time to make clear decisions at the end of the declutter.
- The goal of a digital declutter, however, is not simply to enjoy time away from intrusive technology. During this monthlong process, you must aggressively explore higher-quality activities to fill in the time left vacant by the optional technologies you're avoiding. This period should be one of strenuous activity and experimentation.
- You want to arrive at the end of the declutter having rediscovered the type of activities that generate real satisfaction, enabling you to confidently craft a better life—one in which technology serves only a supporting role for more meaningful ends.

STEP #3: REINTRODUCE TECHNOLOGY

After your thirty-day break comes the final step of the digital declutter: reintroducing optional technologies back into your life. This reintroduction is more demanding than you might imagine.

Some of the participants in my mass declutter experiment treated the process only as a classical digital detox—reintroducing *all* their optional technologies when the declutter ended. This is a mistake. The goal of this final step is to start from a blank slate and only let back into your life technology that passes your strict minimalist standards. It's the care you take here that will determine whether this process sparks lasting change in your life.

With this in mind, for each optional technology that you're considering reintroducing into your life, you must first ask: Does this technology directly support something that I deeply value? This is the only condition on which you should let one of these tools into your life. The fact that it offers *some* value is irrelevant—the digital minimalist deploys technology to serve the things they find most important in their life, and is happy missing out on everything else. For example, when asking this first question, you might decide that browsing Twitter in search of distraction doesn't support an important value. On the other hand, keeping up with your cousin's baby photos on Instagram does seem to support the importance you place on family.

Once a technology passes this first screening question, it must then face a more difficult standard: Is this technology the *best* way to support this value? We justify many of the technologies that tyrannize our time and attention with some tangential connection to something we care about. The minimalist, by contrast, measures the value of these connections and is unimpressed by all but the most robust. Consider our above example about following your cousin's baby pictures on Instagram. We noted that this activity might be tentatively justified by the fact that you deeply value family. But the relevant follow-up question is whether browsing Instagram photos is the *best* way to support this value. On some reflection, the answer is probably no. Something as simple as actually calling this cousin once a month or so would probably prove significantly more effective in maintaining this bond.

If a technology makes it through both of these screening questions, there's one last question you must ask yourself before it's allowed back into your life: How am I going to use this technology going forward to maximize its value and minimize its harms? A point I explore in part 2 is that many attention economy companies want you to think about their services in a binary way: either you use it, or you don't. This allows them to entice you into their ecosystem with some feature you find important, and then, once you're a "user," deploy attention engineering to overwhelm you with integrated options, trying to keep you engaging with their service well beyond your original purpose.

Digital minimalists combat this by maintaining standard

operating procedures that dictate when and how they use the digital tools in their lives. They would never simply say, "I use Facebook because it helps my social life." They would instead declare something more specific, such as: "I check Facebook each Saturday on my computer to see what my close friends and family are up to; I don't have the app on my phone; I culled my list of friends down to just meaningful relationships."

Pulling together these pieces, here's a summary of this minimalist screening process.

The Minimalist Technology Screen

To allow an optional technology back into your life at the end of the digital declutter, it must:

1. Serve something you deeply value (offering *some* benefit is not enough).
2. Be the *best* way to use technology to serve this value (if it's not, replace it with something better).
3. Have a role in your life that is constrained with a standard operating procedure that specifies *when* and *how* you use it.

You can apply this screen to any new technology that you're considering adopting. When you deploy it at the end of a digital declutter, however, it becomes particularly effective, as the preceding break from these technologies provides you with clarity on your values and confidence that your life doesn't actually require that you slavishly stick with the digital status quo. If you're like many of the participants in my mass declutter experiment, the role of technology in your life will look quite different after navigating the reintroduction step with the above screening process.

An electrical engineer named De, for example, was surprised to discover during his digital declutter how addicted he had become to checking news online, and how anxious it was making him—especially politically charged articles. "I dumped all news during [my declutter] and loved it," he told me. "Ignorance is truly bliss sometimes." When the declutter concluded, he recognized that a complete news blackout was not sustainable but also recognized that subscribing to dozens of email newsletters and compulsively checking breaking news sites was not the *best* way to satisfy his need to be informed. He now checks AllSides.com once a day—a news site that covers the top stories, but for each story it neutrally links to three articles: one from a source associated with the political left, one from the right, and one from the center. This format has a way of defusing the aura of emotional charge that permeates a lot of our current political coverage, allowing De to stay up to speed without the spike in anxiety.

Kate solved this same problem by replacing reading news with listening to a news roundup podcast each morning— keeping her informed without providing her the opportunity to mindlessly browse. Mike, by contrast, found it effective to replace all online news with an older technology: the radio.

He discovered that putting on NPR in the background while working on manual tasks kept him sufficiently up to speed and saved him from many of the worst features of internet news. Ramel also embraced an older technology: instead of checking social media feeds to stay up to date, he now has a newspaper delivered to his NYU dorm.

Perhaps predictably, many of the participants in my mass declutter experiment ended up abandoning the social media services that used to take up so much of their time. These services have a way of entering your life through cultural pressure and vague value propositions, so they tend not to hold up well when subjected to the rigor of the screen described above. It was also common, however, for participants to reintroduce social media in a limited manner to serve specific purposes. In these cases, they were often quite rigorous in taming the services with strict operating procedures.

Marianna, for example, now restricts herself to checking her remaining social media services once a week, during the weekend. A sales engineer named Enrique told me that "Twitter is what caused me the most harm," so he also restricted himself to checking his feed only once a week, on the weekend. Ramel and Tarald decided it was sufficient to take their remaining social media apps off of their phones. The extra difficulty involved in accessing these services through a web browser on their desktop computers seemed sufficient to concentrate their use to only the most important purposes.

An interesting experience shared by some participants was that they eagerly returned to their optional technologies only to learn they had lost their taste for them. Here, for example, is how Kate described this experience to me:

> The day the declutter was over, I raced back to Facebook, to my old blogs, to Discord, gleeful and ready to dive back in—and then, after about thirty minutes of aimless browsing, I kind of looked up and thought . . . why am I doing this? This is . . . boring? This isn't bringing me any kind of happiness. It took a declutter for me to notice that these technologies aren't actually adding anything to my life.

She hasn't used those services since.

Several participants discovered that eliminating the point-and-click relationship maintenance enabled by social media requires that you introduce alternative systems for connecting with your friends. A digital advertiser named Ilona, for example, set up a regular schedule for calling and texting her friends—which supported her most serious relationships at the cost of some of the more lightweight touches many have come to expect. "In the end, I just accepted the fact that I would miss some events in their lives, but that this was worthwhile for the mental energy it would save me to not be on social media."

Other participants settled on unusual operating procedures during the reintroduction process. Abby, a Londoner who works in the travel industry, removed the web browser from her phone—a nontrivial hack. "I figured I didn't need to know the answer to everything instantly," she told me. She then bought an old-fashioned notebook to jot down ideas when she's bored on the tube. Caleb set a curfew for his phone: he can't use it between the hours of 9 p.m. and 7 a.m., while a computer engineer named Ron gives himself a quota of only two websites he's allowed to regularly check—a big improvement over the forty or more sites he used to cycle through. Rebecca transformed her daily experience by buying a watch. This might sound trivial to older readers, but to a nineteen-year-old like Rebecca, this was an intentional act. "I estimate that around 75 percent of the time I got sucked down a rabbit hole of un-productivity was due to me checking my phone for the time."

■ ■ ■

To summarize the main points about this step:

- Your monthlong break from optional technologies resets your digital life. You can now rebuild it from scratch in a much more intentional and minimalist manner. To do so, apply a three-step technology screen to each optional technology you're thinking about reintroducing.
- This process will help you cultivate a digital life in which new technologies serve your deeply held values as opposed to subverting them without your permission. It is in this careful reintroduction that you make the intentional decisions that will define you as a digital minimalist.

PART 2

Practices

Spend Time Alone

WHEN SOLITUDE SAVED THE NATION

When you drive north from the National Mall in Washington, DC, on Seventh Street, your route begins among condo buildings and monumental stone architecture. After around two miles it shifts to the brick row houses and crowded restaurants of the close-in city neighborhoods: Shaw, then Columbia Heights, and then, finally, Petworth. Many of the commuters who follow this route up through Petworth don't realize that just a couple of blocks to the east, hidden behind a concrete perimeter wall and a gatehouse manned by a soldier, is a pocket of calm.

The property is the Armed Forces Retirement Home, which has been located here in the heights overlooking downtown DC since 1851, when, under pressure from Congress, the federal government bought the land from banker George Riggs to build a home for disabled veterans of the country's recent wars. In the nineteenth century, the Soldiers' Home (as it was originally called) was surrounded by countryside. Today the city sprawls well beyond the property, but when you pull through its main gates, as I did on an unseasonably warm fall afternoon while researching this book, its ability to provide a sense of escape remained intact. As I drove onto the grounds, the noise of the city diminished: there were green lawns, old trees, chirping birds, and the laughter of children from a nearby charter school playing on a playground. As I turned into a visitors' parking lot, I caught my first glimpse of what I had come to see, the sprawling, thirty-five-room Gothic Revival–style "cottage" originally built by George Riggs and recently restored to recapture how it would have appeared in the 1860s.

This cottage is now a National Historic Site because it once played host to a famous visitor: during each summer and early fall of 1862, 1863, and 1864, President Abraham Lincoln resided there, commuting back and forth to the White House on horseback. But this site is more than just a place where an important president stayed. A growing amount of research suggests that the time and space for quiet reflection the cottage enabled may have played a key role in helping Lincoln make sense of the traumas of the Civil War and tackle the hard decisions he faced.

It was this idea, that something as simple as silence might have shaped our country's history, that brought me to Lincoln's cottage that fall afternoon to find out more.

■ ■ ■

To understand Lincoln's drive to escape the White House, you must imagine what life was like for this untested, one-term congressman thrust unexpectedly into command during our country's most trying period to date. Immediately after Lincoln's inauguration, the day he gave his heady "better angels of our nature" address and attempted to convince the splintering union that it could endure, Lincoln was launched into a whirlwind of duty and distraction. "This president had absolutely no honeymoon," writes historian William Lee Miller. "[He] had no calm first days in which he could settle into the presidential office . . . and think his way toward what he wanted to do by careful steps." Instead, as Miller colorfully puts it: "He was slapped in the face the first business minute of his presidency by the necessity of decision." Miller is not exaggerating. As Lincoln later conveyed to his friend Senator Orville Browning: "The first thing that was handed to me after I entered this room, when I came from the Inauguration, was the letter from Maj. Anderson saying that their provisions would be exhausted." Major Anderson was the commander of the besieged Fort Sumter in Charleston—the fulcrum on which the threat of a looming civil war then rested. The decision on whether to evacuate or defend Sumter was just the first of an avalanche of similar crises that Lincoln faced daily as the executive of a union sliding toward dissolution.

The gravity of these times was not enough to free Lincoln

from other less weighty obligations that relentlessly claimed most of the remaining scraps of his schedule. "Virtually from Lincoln's first day in office," writes Lincoln scholar Harold Holzer, "a crush of visitors besieged the White House stairways and corridors, climbed through windows at levees, and camped outside Lincoln's office door." These visitors arrived to petition for jobs or other personal favors, and included friends and more than a few relatives of Mary Lincoln. The White House Historical Association preserves an engraving in their archives, originally published in a newspaper a month after Lincoln's inauguration, that succinctly captures this reality. It shows a crowd of two dozen top-hatted men milling right outside the doors to the room where Lincoln was meeting with his cabinet. They were there, the caption explains, to aggressively seek employment as soon as the president emerged.

Even though Lincoln eventually attempted to better organize these visitors—making them take turns, "as if waiting to be shaved at a barber's shop," Lincoln joked—dealing with the public remained, as Holzer summarizes, "the biggest drain on the president's time and energy." Against this backdrop of bustle, Lincoln's decision to spend almost half the year escaping the White House, setting out each night to make the long horseback commute to the quiet cottage at the Soldiers' Home, makes sense. The cottage provided Lincoln something we now see would have been almost impossible to obtain in the White House: time and space to think.

Mary and the president's son Tad lived with Lincoln at the

cottage (their older son, Robert, was away at college), but they frequently traveled, so the president often had the sprawling house to himself. To be sure, Lincoln was never literally by himself at the Soldiers' Home: in addition to his household staff, two companies of the 150th Pennsylvania volunteers were camped on the lawn to provide protection. But what made his time at the cottage special was the lack of people demanding his attention: even when he wasn't technically alone, Lincoln was able to be alone with his thoughts.

We know that Lincoln took advantage of this quiet to think because many accounts of people coming to visit Lincoln at the cottage specifically mention that their arrival interrupted his solitude. A letter written by a Treasury employee named John French, for example, describes the following scene when he arrived unannounced with his friend Colonel Scott during the early darkness of a summer evening:

> The servant who answered the bell led the way into the little parlor, where, in the gloaming, entirely alone, sat Mr. Lincoln. [Having] thrown off coat and shoes, and with a large palm-leaf fan in his hand . . . he reposed in a broad chair, one leg hanging over its arm, he seemed to be in deep thought.

Lincoln's commute through the countryside between the capital and his cottage also provided time for him to think. We know Lincoln valued this source of solitude, as he would occasionally sneak out to begin his ride back to the capital without the cavalry company assigned to protect him. This was not a decision made lightly, as the military had previously uncovered a Confederate plot to assassinate Lincoln on this route, and the president was shot at on at least one occasion during the ride.

This time to reflect likely refined Lincoln's responses to key events during his presidency. Folklore, for example, describes Lincoln scribbling the Gettysburg Address on the train ride to deliver his famed speech. This was not, however, Lincoln's usual process: he typically worked on drafts for weeks leading up to important events. As Erin Carlson Mast, the executive director of the nonprofit that oversees the cottage, explained to me during my visit, during the weeks leading up to the Gettysburg Address, Lincoln . . .

> was here at the cottage, often walking alone at night in the military cemetery. He didn't keep a diary, so we don't know his innermost thoughts, but we know he was here, encountering the human cost of war, right before he wrote those memorable lines.

The cottage also provided the setting where Lincoln wrestled with the Emancipation Proclamation. Both the necessity to free southern slaves and the form that this emancipation should take were complicated questions that vexed the Lincoln administration—especially at a time when they were terrified of losing the border slave states to the Confederacy. Lincoln invited visitors like Senator Orville Browning to the cottage to discuss the relevant issues. The president would also famously record his ideas on scraps of paper that he would sometimes store in the lining of his top hat as he wandered the grounds.

Lincoln eventually wrote the initial drafts of the proclamation at the cottage. When I toured the house, I saw the desk where Lincoln first penned those important words. It sits in his high-ceilinged bedroom, between two tall windows that overlook the back lawn. When Lincoln sat there, he would have seen the tents of the Union soldiers camped on the grass of the lawn and, a few miles beyond, the dome of the nation's Capitol, which at the time, like the country, was still under construction.

The desk I saw in Lincoln's cottage is a replica, as the original was moved to the Lincoln Bedroom of the White House. This is ironic because almost certainly Lincoln would have struggled much more with this historical task if he had been forced to grapple with it amid the bustle and distraction of his official residence.

■ ■ ■

Lincoln's time alone with his thoughts played a crucial role in his ability to navigate a demanding wartime presidency. We can therefore say, with only mild hyperbole, that in a certain sense, solitude helped save the nation.

The goal of this chapter is to argue that the benefits Lincoln received from his time alone extend beyond historical figures or those similarly faced with major decisions. *Everyone*

benefits from regular doses of solitude, and, equally important, anyone who avoids this state for an extended period of
time will, like Lincoln during his early months in the White
House, suffer. In the pages ahead, I hope to convince you that,
regardless of how you decide to shape your digital ecosystem,
you should follow Lincoln's example and give your brain the
regular doses of quiet it requires to support a monumental life.

THE VALUE OF SOLITUDE

Before we can usefully discuss solitude, we need to better understand what we mean by this term. To aid us in this effort,
we can turn toward an unlikely pair of guides: Raymond
Kethledge and Michael Erwin.

Kethledge is a respected judge serving on the United States
Court of Appeals for the Sixth Circuit,* and Erwin is a former
army officer who served in both Iraq and Afghanistan. They
first met in 2009, when Erwin was stationed in Ann Arbor to
study toward a graduate degree. Though Kethledge and Erwin
were separated in both age and life experiences, it didn't take
long for them to recognize a shared interest in the topic of
solitude. Kethledge, it turned out, relies on long periods alone
with his thoughts to write his famously sharp legal opinions,

---

* The name Raymond Kethledge may sound familiar, as in the summer of
2018 he was reported to be one of four names on President Donald
Trump's shortlist for Supreme Court nominee to replace Anthony
Kennedy.

often working at a simple pine desk in a barely renovated barn with no internet connection. "I get an extra 20 IQ points from being in that office," he explains. Erwin, for his part, used long runs alongside the cornfields of Michigan to work through the difficult emotions he faced on first returning from combat, joking that "running is cheaper than therapy."

Soon after their initial meeting, Kethledge and Erwin decided to co-write a book on the topic of solitude. It took them seven years, but their efforts culminated in the 2017 release of *Lead Yourself First*. The book summarizes, with the tight logic you expect from a federal judge and former military officer, the authors' case for the importance of being alone with your thoughts. Before outlining their case, however, the authors start with what is arguably one of their most valuable contributions, a precise definition of *solitude*. Many people mistakenly associate this term with physical separation—requiring, perhaps, that you hike to a remote cabin miles from another human being. This flawed definition introduces a standard of isolation that can be impractical for most to satisfy on any sort of a regular basis. As Kethledge and Erwin explain, however, solitude is about what's happening in your brain, not the environment around you. Accordingly, they define it to be a subjective state in which your mind is free from input from other minds.

You can enjoy solitude in a crowded coffee shop, on a subway car, or, as President Lincoln discovered at his cottage, while sharing your lawn with two companies of Union soldiers, so long as your mind is left to grapple only with its own thoughts. On the other hand, solitude can be banished in even the quietest setting if you allow input from other minds to intrude. In addition to direct conversation with another person, these inputs can also take the form of reading a book, listening to a podcast, watching TV, or performing just about any activity that might draw your attention to a smartphone screen. Solitude requires you to move past reacting to information created by other people and focus instead on your own thoughts and experiences—wherever you happen to be.

Why is solitude valuable? Kethledge and Erwin detail many benefits, most of which concern the insight and emotional balance that comes from unhurried self-reflection. Of the many case studies they present, one that resonated particularly strongly concerned Martin Luther King Jr. They note that King's involvement in the Montgomery bus boycott started haphazardly—King happened to be the charismatic and well-educated new minister in town when the local chapter of the NAACP decided to take a stand against bus segregation policies. King's subsequent nomination to be the leader of the newly formed Montgomery Improvement Association, which occurred in a church meeting in late 1955, caught King off guard. He agreed only reluctantly, saying, "Well if you think I can render some service, I will."

As the boycott dragged on, pressures increased both on King's leadership and his personal safety. These pressures were particularly intense given the unintentional manner in which King had become involved in the boycott. These forces culminated on January 27, 1956, the night after King was released from his first stint in jail, where he had been locked up as part of an organized campaign of police harassment. King returned home after his wife and young daughter had gone to sleep, and realized that the time had come for him to clarify what he was about. Sitting alone with his thoughts, holding a cup of coffee at his kitchen table, King prayed and reflected. He embraced the solitude needed to make sense of the demands placed upon him, and in this space he found the answer that would provide him the courage needed for what was ahead:

> And it seemed at that moment that I could hear an inner voice saying to me, "Martin Luther, stand up for righteousness. Stand up for justice. Stand up for truth."

Biographer David Garrow later described this event as "the most important night of [King's] life."

■ ■ ■

Erwin and Kethledge are not, of course, the first commentators to notice the importance of solitude. Its benefits have been explored since at least the early years of the Enlightenment.* "All of humanity's problems stem from man's inability

---

* Solitude has been studied in various guises in a religious context back through antiquity, where it has long served important purposes in helping connect to the divine and sharpen moral intuition. I pick up this thread relatively late in the history of civilization mainly for concision's sake.

to sit quietly in a room alone," Blaise Pascal famously wrote in the late seventeenth century. Half a century later, and an ocean away, Benjamin Franklin took up the subject in his journal: "I have read abundance of fine things on the subject of solitude. . . . I acknowledge solitude an agreeable refreshment to a busy mind."*

The academy was late to recognize the importance of time alone with your own thoughts. In 1988, the noted English psychiatrist Anthony Storr helped correct this omission with his seminal book, *Solitude: A Return to the Self*. As Storr noted, by the 1980s, psychoanalysis had become obsessed with the importance of intimate personal relationships, identifying them as the most important source of human happiness. But Storr's study of history didn't seem to support this hypothesis. He opens his 1988 book with the following quote from Edward Gibbon: "Conversation enriches the understanding, but solitude is the school of genius." He then boldly writes: "Gibbon is surely right."

Edward Gibbon lived a solitary life, but not only did he produce wildly influential work, he also seemed perfectly happy. Storr notes that the need to spend a great deal of time alone was common among "the majority of poets, novelists, and composers." He lists Descartes, Newton, Locke, Pascal, Spinoza,

---

* It's worth noting that Franklin followed up this note in praise of solitude by cautioning that spending *too* much time alone is not good for a "sociable being." His exact quip: "Were these thinking people [who value solitude] obliged to always be alone, I am apt to think they would quickly find their very being insupportable to them."

Kant, Leibniz, Schopenhauer, Nietzsche, Kierkegaard, and Wittgenstein as examples of men who never had families or fostered close personal ties, yet still managed to lead remarkable lives. Storr's conclusion is that we're wrong to consider intimate interaction as the sine qua non of human thriving. Solitude can be just as important for both happiness and productivity.

It's hard to ignore the fact that the entirety of Storr's list of remarkable lives, as well as many of the other historical examples cited above, focus on men. As Virginia Woolf argued in her 1929 feminist manifesto, *A Room of One's Own*, this imbalance should not come as a surprise. Woolf would agree with Storr that solitude is a prerequisite for original and creative thought, but she would then add that women had been systematically denied both the literal and figurative room of their own in which to cultivate this state. To Woolf, in other words, solitude is not a pleasant diversion, but instead a form of liberation from the cognitive oppression that results in its absence.

In Woolf's time, women were denied this liberation by a patriarchal society. In our time, this oppression is increasingly self-inflicted by our preference for the distraction of the digital screen. This is the theme taken up by a Canadian social critic named Michael Harris in his 2017 book, also titled *Solitude*. Harris is concerned that new technologies help create a culture that undermines time alone with your thoughts, noting that "it matters enormously when that resource is under attack." His survey of the relevant literature then points to three crucial benefits provided by solitude: "new ideas; an understanding of the self; and closeness to others."

We've already discussed the first two benefits from this list, but the third is somewhat unexpected and therefore worth briefly unpacking—especially considering how relevant it will become when we later explore solitude's tension with the benefits of connectivity. Harris argues, perhaps counterintuitively, that "the ability to be alone . . . is anything but a rejection of close bonds," and can instead affirm them. Calmly experiencing separation, he argues, builds your appreciation for interpersonal connections when they do occur. Harris is not the first to note this connection. The poet and essayist May Sarton explored the strangeness of this point in a 1972 diary entry, writing:

> I am here alone for the first time in weeks, to take up my "real" life again at last. That is what is strange— that friends, even passionate love, are not my real life unless there is time alone in which to explore and to discover what is happening or has happened. Without the interruptions, nourishing and maddening, this life would become arid. Yet I taste it fully only when I am alone . . .

Wendell Berry summarized this point more succinctly when he wrote: "We enter solitude, in which also we lose loneliness."

■ ■ ■

Examples similar to those given above are voluminous and
point to a clear conclusion: regular doses of solitude, mixed
in with our default mode of sociality, are necessary to flour-
ish as a human being. It's more urgent now than ever that
we recognize this fact, because, as I'll argue next, for the
first time in human history solitude is starting to fade away
altogether.

SOLITUDE DEPRIVATION

The concern that modernity is at odds with solitude is not
new. Writing in the 1980s, Anthony Storr complained that
"contemporary Western culture makes the peace of solitude
difficult to attain." He pointed to Muzak and the recent invention of the "car telephone" as the latest evidence of this en-
croachment of noise into all parts of our lives. Over a hundred
years earlier, Thoreau demonstrated similar concern, famously
writing in *Walden* that "we are in great haste to construct a
magnetic telegraph from Maine to Texas; but Maine and
Texas, it may be, have nothing important to communicate."
The question before us, then, is whether our current moment
offers a *new* threat to solitude that is somehow more pressing
than those that commentators have bemoaned for decades. I
argue that the answer is a definitive yes.

To understand my concern, the right place to start is the iPod revolution that occurred in the first years of the twenty-first century. We had portable music before the iPod, most commonly in the form of the Sony Walkman and Discman (and their competitors), but these devices played only a restricted role in most people's lives—something you used to entertain yourself while exercising, or in the back seat of a car on a long family road trip. If you stood on a busy city street corner in the early 1990s, you would not see too many people sporting black foam Sony earphones on their way to work.

By the early 2000s, however, if you stood on that same street corner, white earbuds would be near ubiquitous. The iPod succeeded not just by selling lots of units, but also by changing the culture surrounding portable music. It became common, especially among younger generations, to allow your iPod to provide a musical backdrop to your *entire* day—putting the earbuds in as you walk out the door and taking them off only when you couldn't avoid having to talk to another human.

To put this in context, previous technologies that threatened solitude, from Thoreau's telegraph to Storr's car phone, introduced new ways to occasionally interrupt time alone with your thoughts, whereas the iPod provided for the first time the ability to be *continuously* distracted from your own mind. The farmer in Thoreau's time might leave the quiet fireside to walk to town and check the evening telegraph dispatches, fragmenting a moment of solitude, but there was no way that this technology could offer continuous distraction to this same farmer as he went about his day. The iPod was pushing us toward a newly alienated phase in our relationship with our own minds.

This transformation started by the iPod, however, didn't reach its full potential until the release of its successor, the iPhone, or, more generally, the spread of modern internet-connected smartphones in the second decade of the twenty-first century. Even though iPods became ubiquitous, there were still moments in which it was either too much trouble to slip in the earbuds (think: waiting to be called into a meeting), or it might be socially awkward to do so (think: sitting bored during a slow hymn at a church service). The smartphone provided a new technique to banish these remaining slivers of solitude: *the quick glance.* At the slightest hint of boredom, you can now surreptitiously glance at any number of apps or mobile-adapted websites that have been optimized to provide you an immediate and satisfying dose of input from other minds.

It's now possible to completely banish solitude from your life. Thoreau and Storr worried about people enjoying less solitude. We must now wonder if people might forget this state of being altogether.

■ ■ ■

Part of what complicates discussions of waning solitude in the smartphone age is that it's easy to underestimate the severity of this phenomenon. While many people admit that they use their phones more than they probably should, they often don't realize the full magnitude of this technology's impact. The

NYU professor Adam Alter, whom I introduced earlier in this book, details a typical story of such underestimation in *Irresistible*. While researching his book, Alter decided to measure his own smartphone use. To do so, he downloaded an app called Moment, which tracks how often and how long you look at your screen each day. Before activating the app, Alter estimated that he probably checks his phone around ten times a day for a total of about an hour of screen time.

A month later, Moment provided Alter the truth: on average, he was picking up his phone forty times per day and spending around a total of three hours looking at his screen. Surprised, Alter contacted Kevin Holesh, the app developer behind Moment. As Holesh revealed, Alter is not an outlier. In fact, he's remarkably typical: the average Moment user spends right around three hours a day looking at their smartphone screen, with only 12 percent spending less than an hour. The average Moment user picks up their phone thirty-nine times a day.

As Holesh reminds Alter, these numbers probably skew low, as the people who download an app like Moment are people who are already careful about their phone use. "There are millions of smartphone users who are oblivious or just don't care enough to track their usage," Alter concludes. "There's a reasonable chance they're spending even more than three hours on their phone each day."

The smartphone usage numbers cited above only count the time spent looking at your screen. When you add in time spent listening to music, audiobooks, and podcasts—none of

SPEND TIME ALONE 103

which are measured by the Moment app—it should become more clear how effective people have become in banishing moments of solitude from their daily experience.

To simplify our discussion, let's give this trend its own name:

> **Solitude Deprivation**
>
> A state in which you spend close to zero time
> alone with your own thoughts and free from
> input from other minds.

As recently as the 1990s, solitude deprivation was difficult to achieve. There were just too many situations in everyday life that forced you to be alone with your thoughts, whether you wanted to or not—waiting in line, crammed into a crowded subway car, walking down the street, working on your yard. Today, as I've just argued, it's become widespread.

The key question, of course, is whether the spread of solitude deprivation should concern us. Tackled abstractly, the answer is not immediately obvious. The idea of being "alone" can seem unappealing, and we've been sold, over the past two decades, the idea that more connectivity is better than less. Surrounding the announcement of his company's 2012 IPO, for example, Mark Zuckerberg triumphantly wrote: "Facebook . . . was built to accomplish a social mission—to make the world more open and connected."

This obsession with connection is clearly overly optimistic, and it's easy to make light of its grandiose ambition, but when solitude deprivation is put into the context of the ideas discussed earlier in this chapter, this prioritization of communication over reflection becomes a source of serious concern. For one thing, when you avoid solitude, you miss out on the positive things it brings you: the ability to clarify hard problems, to regulate your emotions, to build moral courage, and to strengthen relationships. If you suffer from chronic solitude deprivation, therefore, the quality of your life degrades.

Eliminating solitude also introduces new negative repercussions that we're only now beginning to understand. A good way to investigate a behavior's effect is to study a population that pushes the behavior to an extreme. When it comes to constant connectivity, these extremes are readily apparent among young people born after 1995—the first group to enter their preteen years with access to smartphones, tablets, and persistent internet connectivity. As most parents or educators of this generation will attest, their device use is *constant*. (The term *constant* is not hyperbole: a 2015 study by Common Sense Media found that teenagers were consuming media—including text messaging and social networks—*nine hours per day* on average.) This group, therefore, can play the role of a cognitive canary in the coal mine. If persistent solitude deprivation causes problems, we should see them show up here first.

And this is exactly what we find.

My first indication that this hyper-connected generation was suffering came a few years before I started writing this book. I was chatting with the head of mental health services at a well-known university where I had been invited to speak. This administrator told me that she had begun seeing major shifts in student mental health. Until recently, the mental health center on campus had seen the same mix of teenage issues that have been common for decades: homesickness, eating disorders, some depression, and the occasional case of OCD. Then everything changed. Seemingly overnight the number of students seeking mental health counseling massively expanded, and the standard mix of teenage issues was dominated by something that used to be relatively rare: anxiety.

She told me that everyone seemed to suddenly be suffering from anxiety or anxiety-related disorders. When I asked her what she thought caused the change, she answered without hesitation that it probably had something to do with smartphones. The sudden rise in anxiety-related problems coincided with the first incoming classes of students that were raised on smartphones and social media. She noticed that these new students were constantly and frantically processing and sending messages. It seemed clear that the persistent communication was somehow messing with the students' brain chemistry.

A few years later, this administrator's hunch was validated by San Diego State University psychology professor Jean Twenge, who is one of the world's foremost experts on generational differences in American youth. As Twenge notes in a September 2017 article for the *Atlantic*, she has been studying these trends for over twenty-five years, and they almost always appear and grow gradually. But starting around 2012, she noticed a shift in measurements of teenager emotional states that was anything but gradual:

> The gentle slopes of the line graphs [charting how behavioral traits change with birth year] became steep mountains and sheer cliffs, and many of the distinctive characteristics of the Millennial generation began to disappear. In all my analyses of generational data—some reaching back to the 1930s—I had never seen anything like it.

Young people born between 1995 and 2012, a group Twenge calls "iGen," exhibited remarkable differences as compared to the Millennials that preceded them. One of the biggest and most troubling changes was iGen's psychological health. "Rates of teen depression and suicide have skyrocketed," Twenge writes, with much of this seemingly due to a massive increase in anxiety disorders. "It's not an exaggeration to describe iGen as being on the brink of the worst mental-health crisis in decades."

What instigated these changes? Twenge agrees with the intuition of the university mental health administrator when she notes that these shifts in mental health correspond "exactly" to the moment when American smartphone ownership became ubiquitous. The defining trait of iGen, she explains, is that they grew up with iPhones and social media, and don't remember a time before constant access to the internet. They're paying a price for this distinction with their mental health. "Much of this deterioration can be traced to their phones," Twenge concludes.

When journalist Benoit Denizet-Lewis investigated this teen anxiety epidemic in the *New York Times Magazine*, he also discovered that the smartphone kept emerging as a persistent signal among the noise of plausible hypotheses. "Anxious kids certainly existed before Instagram," he writes, "but many of the parents I spoke to worried that their kids' digital habits— round-the-clock responding to texts, posting to social media, obsessively following the filtered exploits of peers—were partly to blame for their children's struggles."

Denizet-Lewis assumed that the teenagers themselves would dismiss this theory as standard parental grumbling, but this is not what happened. "To my surprise, anxious teenagers tended to agree." A college student he interviewed at a residential anxiety treatment center put it well: "Social media is a tool, but it's become this thing that we can't live without that's making us crazy."

As part of his reporting, Denizet-Lewis interviewed Jean Twenge, who made it clear that she didn't set out to implicate the smartphone: "It seemed like too easy an explanation for negative mental-health outcomes in teens," but it ended up the only explanation that fit the timing. Lots of potential culprits, from stressful current events to increased academic pressure, existed before the spike in anxiety that begins around 2011. The only factor that dramatically increased right around the same time as teenage anxiety was the number of young people owning their own smartphones.

"The use of social media and smartphones look culpable for the increase in teen mental-health issues," she told Denizet-Lewis. "It's enough for an arrest—and as we get more data, it might be enough for a conviction." To emphasize the urgency of this investigation, Twenge titled her article in the *Atlantic* with a blunt question: "Have Smartphones Destroyed a Generation?"

Returning to our canary-in-the-coal-mine analogy, the plight of iGen provides a strong warning about the danger of solitude deprivation. When an entire cohort unintentionally eliminated time alone with their thoughts from their lives, their mental health suffered dramatically. On reflection, this makes sense. These teenagers have lost the ability to process and make sense of their emotions, or to reflect on who they are and what really matters, or to build strong relationships, or even to just allow their brains time to power down their critical social circuits, which are not meant to be used constantly, and to redirect that energy to other important cognitive housekeeping tasks. We shouldn't be surprised that these absences lead to malfunctions.

Most adults stop short of the constant connectivity practiced by members of iGen, but if you extrapolate these effects to the somewhat milder forms of solitude deprivation that have become common among many different age groups, the results are still worrisome. As I've learned by interacting with my readers, many have come to accept a background hum of low-grade anxiety that permeates their daily lives. When looking for explanations, they might turn to the latest crisis— be it the recession of 2009 or the contentious election of 2016—or chalk it up to a normal reaction to the stresses of adulthood. But once you begin studying the positive benefits of time alone with your thoughts, and encounter the distressing effects that appear in populations that eliminate this altogether, a simpler explanation emerges: we *need* solitude to thrive as human beings, and in recent years, without even realizing it, we've been systematically reducing this crucial ingredient from our lives.

Simply put, humans are not wired to be constantly wired.

THE CONNECTED CABIN

Assuming you accept my premise that solitude is necessary to thrive as a human being, the natural follow-up question is: How can you find enough of this solitude in the hyperconnected twenty-first century? To answer it, we can draw an unexpected insight from Thoreau's cabin at Walden Pond.

Thoreau's retreat to the woods beyond Concord, Massachusetts, with the intention to live more deliberately, is cited as a classic example of solitude. Thoreau helped spread this conception. His book about the experience, *Walden*, is rich with long passages describing Thoreau alone and observing the slow rhythms of nature. (You'll never think of pond ice the same way again after you read Thoreau's lengthy discussion of how its qualities change throughout the winter.)

In the decades since *Walden*'s release, however, critics have been busy attacking the mythology of Walden as an isolated outpost. Historian W. Barksdale Maynard, to cite one example among many, listed in a 2005 essay the many ways in which Thoreau was anything but isolated during his time at the pond. Thoreau's cabin, it turns out, was not in the woods, but in a clearing near the woods that was in sight of a well-traveled public road. Thoreau was only a thirty-minute walk from his hometown of Concord, where he returned regularly for meals and social calls. Friends and family, for their part, visited him constantly at his cabin, and Walden Pond, far from an untrammeled oasis, was then, as it remains today, a popular destination for tourists seeking a nice walk or swim.

But as Maynard explains, this complicated mixture of solitude and companionship is not a secret Thoreau was trying to hide. It was, in some sense, the whole point. "[Thoreau's] intention was not to inhabit a *wilderness*," he writes, "but to find *wildness* in a suburban setting."

We can substitute *solitude* for *wildness* in this sentence without changing its meaning. Thoreau had no interest in complete disconnection, as the intellectual milieu of mid-nineteenth-century Concord was surprisingly well developed and Thoreau didn't want to completely disengage from this energy. What Thoreau sought in his experiment at Walden was the ability to move back and forth between a state of solitude and a state of connection. He valued time alone with his thoughts—staring at ice—but he also valued companionship and intellectual stimulation. He would have rejected a life of true hermit-style isolation with the same vigor with which he protested the thoughtless consumerism of the early industrial age.

This cycle of solitude and connection is a solution that comes up often when studying people who successfully side-step solitude deprivation; think, for example, of Lincoln spending his summer nights at his cottage before returning to the bustling White House in the morning, or of Raymond Kethledge taking a break from the busy courthouse to clarify his thoughts in a quiet barn. The pianist Glenn Gould once proposed a mathematical formula for this cycle, telling a journalist: "I've always had a sort of intuition that for every hour you spend with other human beings you need X number of hours alone. Now what that X represents I don't really know . . . but it's a substantial ratio."

It's exactly this alternation between regular time alone with your thoughts and regular connection that I propose as the key to avoiding solitude deprivation in a culture that also demands connection. As Thoreau's example emphasizes, there's nothing wrong with connectivity, but if you don't balance it with regular doses of solitude, its benefits will diminish.

To help you realize this cycle in your modern life, this chapter concludes with a small collection of *practices*—each of which offers a specific and effective approach to integrating more solitude into an otherwise connected routine. These practices are not exhaustive nor are they obligatory. Think of

them instead as a look at the varied ways that people have succeeded in creating their own metaphorical cabin by the pond
in an increasingly noisy world.

PRACTICE: LEAVE YOUR
PHONE AT HOME

The Alamo Drafthouse Cinema in Austin, Texas, doesn't
allow you to use phones once the film begins. The glow of
the screen distracts patrons from the cinematic experience,
and the Alamo Drafthouse is the type of place where people
respect cinematic experience. Most movie theaters, of course,
politely ask moviegoers to put away their phones, but this
particular venue takes this prohibition seriously. Here's their
official policy, lifted from their website:

> We have zero tolerance for talking or cell phone use
> of any kind during films. We'll kick you out, promise.
> We've got backup.

This policy is notable in part because it's so exceptional in
the movie business. The standard multiplex has implicitly
given up on the idea that people can make it through a film
without using their phone. Some are even considering formal-
izing this retreat. "You can't tell a 22-year-old to turn off their
cellphone," said the CEO of the AMC theater chain in a 2016
interview with *Variety*. "That's not how they live their life."

He then revealed that the company is considering relaxing their existing (though largely ignored) cell phone bans.

The failed fight against cell phones in movie theaters is a specific consequence of a more general shift that's occurred over the past decade: the transformation of the cell phone from an occasionally useful tool to something we can never be apart from. This rise of *cell phone as vital appendage* is supported by many different explanations. Young people, for example, worry that even temporary disconnection might lead them to miss out on something better they could be doing. Parents worry that their kids won't be able to reach them in an emergency. Travelers need directions and recommendations for places to eat. Workers fear the idea of being both needed and unreachable. And everyone secretly fears being bored.

What's remarkable about these concerns is how recently we started really caring about them. People born before the mid-1980s have strong memories of life without cell phones. All of the concerns listed above still existed in theory, but no one worried much about them. Before I had my driver's license, for example, if I needed someone to pick me up from school after sports practice, I'd use a payphone: sometimes my parents were home, and sometimes I had to leave a message and hope they got it. Getting lost and asking for directions was just a regular part of driving in a new city, and not really a big deal—learning to read maps was one of the first things I did after learning to drive. Parents were comfortable with the idea that when they were out for dinner and a movie, the babysitter had no easy way to reach them in the case of an emergency.

I don't mean to create a false sense of nostalgia for these pre–cell phone times. All of the above scenarios are somewhat improved by better communication tools. But what I do want to emphasize is that most of this improvement is minor. Put another way, in 90 percent of your daily life, the presence of a cell phone either doesn't matter or makes things only slightly more convenient. They're useful, but it's hyperbolic to believe its ubiquitous presence is vital.

This claim can be validated in part by turning to the surprisingly vibrant subculture of people who go extended periods without cellular communication. We know about this group because many of them publish essays describing their experience. If you read enough of these dispatches, a common theme emerges: life without a cell phone is occasionally annoying, but it's much less debilitating than you might expect.

A young woman named Hope King, for example, ended up spending a little over four months without a phone after her iPhone was stolen at a clothing store. She could have replaced it right away, but delaying this decision struck her at the time as an act of symbolic defiance against the thief—a perhaps misguided, but good-intentioned way of saying, "See, you didn't hurt me." In an article she wrote about her experience, King listed several "nuisances" of life without a phone, including the need to look up maps in advance before heading to a new destination, and the slightly increased complexity of talking with her family (which she did over Skype on her laptop). She also experienced a small number of major annoyances, such as the time she was stuck in the back of a taxi, running late for a meeting with her boss, desperately hoping to snag a Wi-Fi signal from a nearby Starbucks on her iPad so she could send him a note. But for the most part, the experience was less drastic than she feared. Indeed, as she writes, some things that concerned her about post–cell phone life "were surprisingly easy," and when she was finally forced to buy a new phone (a new job required it), she actually felt anxious about the return to constant connection.

The purpose of these observations is to underscore the following point: the urgency we feel to always have a phone with us is exaggerated. To live permanently without these devices would be needlessly annoying, but to regularly spend a few hours away from them should give you no pause. It's important that I convince you of this reality, as spending more time away from your phone is exactly what I'm going to ask you to do.

■ ■ ■

I argued earlier in this chapter that smartphones are the primary enabler of solitude deprivation. To avoid this condition, therefore, it makes sense to try to spend regular time away from these devices—re-creating the frequent exposure to solitude that until recently was an unavoidable part of daily life. I recommend that you try to spend some time away from your phone *most* days. This time could take many forms, from a quick morning errand to a full evening out, depending on your comfort level.

Succeeding with this strategy requires that you abandon

the belief that not having your phone is a crisis. As I argued above, this belief is new and largely invented, but it can still take some practice before you fully accept its truth. If you're struggling at first, a useful compromise is to bring your phone where you're going, but then leave it in your car's glove compartment. This way, if there's an emergency that requires connection, you can always go retrieve your device, but it's not *right there* with you where it can destroy solitude at a moment's notice. If you're not driving but out with someone else, it can work just as well to have them hold your phone for you (assuming you can convince them to do so)—as before, you have emergency access, but not easy access.

To emphasize what I hope is clear, this practice is not about getting rid of your phone—most of the time, you'll have your phone with you and enjoy all of its conveniences. It does aim, however, to convince you that it's completely reasonable to live a life in which you sometimes have a phone with you, and sometimes do not. Indeed, not only is this lifestyle reasonable, but it represents a small behavior tweak that can reap large benefits by protecting you from the worst effects of solitude deprivation.

PRACTICE: TAKE LONG WALKS

In 1889, as Friedrich Nietzsche's fame began to spread, he published a brief introduction to his philosophy. It was called *Twilight of the Idols*, and it took him only two weeks to write.

Early in the book is a chapter that contains aphorisms on topics that interested Nietzsche. It's in this chapter, more specifically in maxim 34, that we find the following strong claim: "Only thoughts reached by walking have value." To underscore his esteem for walking, Nietzsche also notes: "The sedentary life is the very sin against the Holy Spirit."

Nietzsche was speaking from personal experience. As the French philosopher Frédéric Gros elaborates in his 2009 book on the intersection of walking and philosophy, Nietzsche, by 1889, was concluding a wildly productive decade in which he rebounded from failing health and wrote some of his greatest books. This period began ten years earlier, when Nietzsche was forced by recurring migraines, among other maladies, to leave his position as a university professor. He submitted his resignation in May 1879 and later that summer found himself in a small village on the Upper Engadine slopes. In the decade that stretched between his resignation and the publication of *Twilight of the Idols*, Nietzsche survived on a series of small grants that provided enough funds for modest lodging and the ability to take the train back and forth between the mountains (where he would escape the summer heat) and the sea (where he would escape the winter cold).

It was during this period, when Nietzsche found himself surrounded by some of Europe's most scenic trails, that "he became the peerless walker of legend." As Gros recounts, during his first summer on the Upper Engadine, Nietzsche began to walk up to eight hours a day. During these walks he would think, eventually filling six small notebooks with the prose that became *The Wanderer and His Shadow*, the first of many influential books he wrote during a decade powered by ambulation.

Nietzsche, of course, is not the only historical figure to use walking to support a contemplative life. In his book, Gros also points to the example of the French poet Arthur Rimbaud, a restless soul who set off on many long pilgrimages on foot, often short of money but rich in passion, or Jean-Jacques Rousseau, who once wrote: "I never do anything but when walking, the countryside is my study." About Rousseau, Gros adds: "The mere sight of a desk and a chair was enough to make him feel sick."

The value of walking also suffuses American culture. Wendell Berry, another proponent of strolling, used long outings through the fields and forests of his rural Kentucky to clarify his pastoral values. As he once wrote:

> As I walk, I am always reminded of the slow, patient
> building of soil in the woods. And I am reminded of
> the events and companions of my life—for my walks,
> after so long, are cultural events.

Berry was likely inspired by Thoreau, who is arguably America's most strident booster of walking. In his famed Lyceum lecture, which was posthumously published in the *Atlantic Monthly* under the title of "Walking," Thoreau labels this activity a "noble art," clarifying: "The walking of which I speak has nothing in it akin to taking exercise . . . but is itself the enterprise and adventure of the day."

■ ■ ■

These historical walkers embraced the activity for different reasons. Nietzsche regained his health and found an original philosophical voice. Berry formalized his intuitive nostalgia. Thoreau found a connection to nature he thought fundamental to a thriving human life. These different reasons, however, are all served by the same key property of walking: it's a fantastic source of solitude. It's important here to remember our technical definition of solitude as freedom from input from other minds, as it's exactly this absence of reaction to the clatter of civilization that supports all of these benefits. Nietzsche emphasized this point when he contrasted the originality of his walk-stimulated ideas with those produced by the bookish scholar locked in a library reacting only to other people's work. "We do not belong," he wrote, "to those who have ideas only among books, when stimulated by books."

Motivated by these historical lessons, we too should embrace walking as a high-quality source of solitude. In doing so, we should heed Thoreau's warning that we're not talking about a short jaunt for a little exercise, but honest-to-goodness, deep-in-the-woods, Nietzsche-on-the-slope-of-a-mountain-style long journeys—these are the grist of productive aloneness.

I've long embraced this philosophy. When I was a postdoc at MIT, my wife and I rented a tiny apartment on Beacon Hill, about a mile's walk across the Longfellow Bridge to the east side of campus where I worked. I made this walk every day, regardless of the weather. I would sometimes meet my wife after work on the banks of the Charles River. If I got there early, I would read. It was on these riverbanks that, appropriately enough, I first discovered the writings of Thoreau and Emerson.

Living, as I now do, in Takoma Park, Maryland, a small town inside the Washington, DC, beltway, I can no longer make long daily walks by a river part of my commute. One of the features that attracted me to this town, however, is its extensive sidewalks shaded by a well-maintained tree canopy. I'm quickly gaining a reputation as that *odd professor* who seems to be constantly wandering up and down the Takoma Park streets.

I use these walks for multiple purposes. The most common activities include trying to make progress on a professional problem (such as a math proof for my work as a computer scientist or a chapter outline for a book) and self-reflection on some particular aspect of my life that I think needs more attention. I sometimes go on what I call "gratitude walks," where I just enjoy particularly good weather, or take in a neighborhood I like, or, if I'm in the middle of a particularly busy or stressful period, try to generate a sense of anticipation for a better season to come. I sometimes start a walk with the intent of tackling one of these goals, and then soon discover my mind has other ideas about what really needs attention. In such instances, I try to defer to my cognitive inclinations, and remind myself how hard it would be to pick up these signals amid the noise that dominates in the absence of solitude.

In short, I would be lost without my walks because they've become one of my best sources of solitude. This practice proposes that you'll find similar benefits by spending more time alone on your feet. The details of this practice are simple: On a regular basis, go for long walks, preferably somewhere scenic. Take these walks alone, which means not just by yourself, but also, if possible, without your phone. If you're wearing headphones, or monitoring a text message chain, or, God forbid, narrating the stroll on Instagram—you're not really walking, and therefore you're not going to experience this practice's greatest benefits. If you cannot abandon your phone for logistical reasons, then put it at the bottom of a backpack so you can use it in an emergency but cannot easily extract it at the first hint of boredom. (If you're worried about not having your phone, see the discussion on this topic in the preceding practice.)

The hardest part of this habit is making the time. In my experience, you'll probably have to invest effort to clear the necessary hours from your schedule—they're unlikely to arise naturally. This might mean, for example, scheduling workday walks on your calendar well in advance (they're a great way to start or end a day), or negotiating with your family some times in the evening or on the weekend when you're going to hit the trail. It also helps if you learn to broaden your definition of "good weather." You can walk on cold days, or when it's snowing, or even during light rain (during my MIT commutes I learned the value of good rain pants). I once even took my dog for a short walk while a hurricane worked its way past Washington, DC, though, in retrospect, this was probably not a smart decision.

These efforts are hard, but the rewards are big. I'm quite simply happier and more productive—by noticeably large factors—when I'm walking regularly. Many others, both today and historically, enjoy the same benefits that come from this substantial injection of solitude into an otherwise hectic life.

Thoreau once wrote:

> I think that I cannot preserve my health and spirits, unless I spend four hours a day at least—and it is commonly more than that—sauntering through the woods and over the hills and fields, absolutely free from all worldly engagements.

Most of us will never meet Thoreau's ambitious commitment to ambulation. But if we remain inspired by his vision, and try to spend as much time as is reasonable on foot and engaging in the "noble art" of walking, we too will experience success in preserving our health and spirits.

PRACTICE: WRITE LETTERS
TO YOURSELF

I have a stack of twelve black, pocket-size Moleskine notebooks on the top shelf of a bookcase in my home office. A thirteenth notebook is currently in my work bag. Given that I bought my first Moleskine in the summer of 2004, and I'm writing these words in the early fall of 2017, this works out to about one notebook per year.

My use of these journals has evolved over time. My very first entry was made on August 7, 2004, in the very first Moleskine I owned. I bought this notebook at the MIT Coop bookstore soon after my arrival in Cambridge to start my life as a graduate student. Its first entry is therefore titled, appropriately enough, "MIT," and it lists some ideas for research projects. The early entries in this first notebook are mainly focused on professional topics. In addition to graduate student issues, it also includes quite a few notes about marketing my first book, *How to Win at College*, which was published in early 2005. These entries are interesting today mainly for their humorously dated cultural references (one such entry solemnly declares, "take a page from [Howard] Dean's campaign: empower people," while another—and I swear I'm not making this up—references both UGG boots and the hit early 2000s reality show *The Osbournes*).

In early 2007, however, the content of my notebooks broadens from a narrow focus on professional projects to also include reflections and ideas about my life more generally. One entry around this period is titled "5 things to focus on this semester," while another details some thoughts on "blank page productivity," an organizational system I was experimenting with at the time. The fall of 2008 sees a more significant shift toward deeper introspection with an entry titled "Better," which lays out a vision for both my professional and personal life. It ends with the earnest request to "accept only excellence from myself."

In December of that year, I wrote an entry titled "The Plan," underneath which I put a list of my values in life, falling under the categories of "relationships," "virtues," and "qualities." I still remember writing this entry on my bed in my fourth-floor walk-up outside Harvard Square. I had just recently come back from sitting shiva with a friend who had lost a parent, and getting a grip on what mattered to me suddenly seemed important. This entry also gets credit for instigating a habit where every time I started a new Moleskine notebook, I would begin by transcribing my current list of values, underneath the heading "The Plan," in the notebook's first pages.

The 2010 entries are particularly interesting, as they contain the seeds of the ideas that grew into my past three books: *So Good They Can't Ignore You*, *Deep Work*, and the title you're currently reading. When I recently reread these notebooks, I was surprised to recall how far my thinking had already developed on issues like the danger of passion in career planning, the power of specialized craftsmanship in an age of general-purpose computing, and, presciently, the appeal of a new brand of technology-focused minimalism—which I was calling "Simplicity 2.0" at the time.

My first child was born in late 2012. Not surprisingly, the 2013 notebook is filled with reflections and urgent plans for figuring out how to be a father. My most recent notebook entries focus quite a bit on trying to clarify the years ahead now that I've succeeded in becoming a tenured professor and

working author. I might be a couple of introspective notebooks away from figuring this out, but if personal history is a
trusted guide, I'll get there.

■ ■ ■

My Moleskine notebooks are not exactly diaries because I
don't write in them on a regular schedule. If you flip through
their pages, you'll encounter an uneven pacing: sometimes I'll
fill dozens of pages in a single week, while other times many
months might pass without any new notes. The uneventful
year of 2006, during which I was mainly just putting my head
down and trying to stay ahead of my graduate coursework, has
no entries at all.

These notebooks play a different role: they provide me a
way to write a letter to myself when encountering a complicated decision, or a hard emotion, or a surge of inspiration. By
the time I'm done composing my thoughts in the structured
form demanded by written prose, I've often gained clarity. I
do make a habit of regularly reviewing these entries, but this
habit is often superfluous. It's the act of writing itself that already yields the bulk of the benefits.

Earlier in this chapter, I introduced Raymond Kethledge
and Michael Erwin's definition of solitude as time spent alone
with your own thoughts and free from inputs from other
minds. Writing a letter to yourself is an excellent mechanism
for generating exactly this type of solitude. It not only frees
you from outside inputs but also provides a conceptual scaf-
folding on which to sort and organize your thinking.

Not surprisingly, I'm not the only person to discover this particular solitude hack. As Kethledge and Erwin report in their book, Dwight Eisenhower leveraged a "practice of thinking by writing" throughout his career to make sense of complicated decisions and tame intense emotions. He was not the only leader to deploy this habit. As mentioned earlier in this chapter, when visiting his cottage at the Soldiers' Home, Abraham Lincoln had a habit of recording thoughts on scraps of paper that he would stick in his hat for safekeeping. (Indeed, the first draft of Lincoln's Emancipation Proclamation was collated, in part, from ideas spanning paper scraps. Inspired by this, the nonprofit that now operates the President Lincoln's Cottage historical site runs a program encouraging young students to do more rigorous original thinking. They call it Lincoln's Hat.)

This practice asks you to embrace this well-validated strategy by making time to write a letter to yourself when faced with demanding or uncertain circumstances. You can follow my lead and keep a special notebook for this purpose, or, like Abraham Lincoln, you can grab a scrap of paper when the need arises. The key is the act of writing itself. This behavior necessarily shifts you into a state of productive solitude—wrenching you away from the appealing digital baubles and addictive content waiting to distract you, and providing you with a structured way to make sense of whatever important things are happening in your life at the moment.

It's a simple practice that's easy to deploy, but it's also incredibly effective.

Don't Click "Like"

THE GREATEST DUEL IN SPORTS

In 2007, ESPN aired what has to be one of the strangest sporting events to ever appear on the channel: the national championship of the USA Rock Paper Scissors League. The title match, which is preserved on YouTube, begins with the play-by-play announcers excitedly describing the two "RPS phenoms" (RPS being short for *rock paper scissors*) that will be competing, declaring with deadpan seriousness that the audience is about to witness the "greatest duel in sports."

The competition is held in a mini boxing ring with a podium in the middle. The first contestant wears glasses and is dressed in khaki pants and a short-sleeve, button-down shirt. He trips on the ropes trying to climb into the ring. His nickname, we're told, is "Land Shark." The second contestant, nicknamed "the Brain," arrives, also dressed in khakis. He makes it into the ring without falling over. "That bodes well," the announcer helpfully explains.

A referee enters and chops his hand over the podium to start the first match. Both players do a three count with their fists before throwing down their signs. The Brain chooses paper while Land Shark chooses scissors. Point to Land Shark! The crowd cheers. A little less than three minutes later, with the score in his favor, Land Shark wins the championship, and the $50,000 grand prize, by smothering the Brain's rock with what the announcers call "the paper heard around the world."

On first encounter, the idea of serious rock paper scissors matches might sound silly. Unlike poker or chess, there doesn't seem to be any room for strategy, which, if true, would make the outcome of a given tournament essentially random. Except this is not what actually happens. During the peak of the league's popularity in the early 2000s, the same high-skilled players kept ending up near the top of tournament rankings, and when accomplished players compete against novices, the role of skill becomes even more pronounced. In a promotional video produced by the national league, a tournament-caliber player who goes by the name Master Roshambollah* challenges strangers to pickup games in a Las Vegas hotel lobby. He wins almost every time.

The explanation for these results is that rock paper scissors, contrary to initial assumption, requires strategy. What sepa-

---

* This moniker is a play on the French name Rochambeau, which is a slang term for rock paper scissors.

rates advanced players like the Brain, Land Shark, and Master Roshambollah from RPS mortals, however, is not a tediously memorized sequence of plays, or statistical wizardry, it's instead their sophisticated grasp of a much broader topic: human psychology.

A strong rock paper scissors player integrates a rich stream of information about their opponent's body language and recent plays to help approximate their opponent's mental state and therefore make an educated guess about the next play. These players will also use subtle movements and phrases to prime their opponent to think about a certain play. The opponent, however, might notice the priming attempt and adjust their play accordingly. Of course, the original player might expect this, and execute a tertiary adjustment, and so on. It should come as no surprise that participants in rock paper scissors tournaments often describe the experience as exhausting.

To see some of these dynamics in action, let's return to the first throw of the 2007 championship match described above. Right before the players begin their three count, the Brain says, "Let's roll." This seems innocuous, but as the play-by-play announcer notes, this is a "subliminal call" for his opponent to play rock (the idea of *rolling* primes the mind to think about *rocks*). After planting this seed to nudge his opponent toward rock, the Brain plays paper. The subliminal strategy, however, backfires. Land Shark notices it and guesses what the Brain is up to, so he plays scissors, beating the Brain's paper and winning the throw.

■ ■ ■

Understanding rock paper scissors champions is important to our purposes because their strategies highlight a foundational endowment shared by every human being on earth: the ability to perform complicated social thinking. To put this ability to use for the narrow purpose of winning an RPS throw requires some game-specific practice, but as I'll elaborate below, most people don't realize the extreme degree to which they perform similarly demanding feats of social navigation and mind reading throughout their normal everyday interactions. Our brains, in many ways, can be understood as sophisticated social computers.

A natural conclusion of this reality is that we should treat with great care any new technology that threatens to disrupt the ways in which we connect and communicate with others. When you mess with something so central to the success of our species, it's easy to create problems.

In the pages ahead, I'll detail the ways in which our brains evolved to crave rich social interaction, and then explore the serious issues caused when we displace this interaction with highly appealing, but much less substantial, electronic pings. I'll then conclude by suggesting a somewhat radical strategy for the digital minimalist looking to sidestep these harms while maintaining the advantages of new communication tools—a strategy that puts these new forms of interaction to work supporting the old.

THE SOCIAL ANIMAL

The idea that humans have a particular affinity for interaction and communication is not new. Aristotle famously noted that "man is by nature a social animal." It wasn't, however, until surprisingly recently in the long sweep of human history that we discovered the biological extent to which this philosophical intuition turns out to be true.

A key moment in this new understanding came in 1997, when a research team from Washington University published a pair of papers in the prestigious *Journal of Cognitive Neuroscience*. During this period, PET scanners, which were originally developed for medical purposes, were migrating into neuroscience research, where they provided researchers the breakthrough ability to observe brain activity. The Washington University team looked at a collection of these new brain imaging studies to investigate a simple question: Are there regions of the brain that are involved in *all* types of brain activity?

As the psychologist Matthew Lieberman summarizes in his 2013 book, *Social*, the results of this initial analysis were "disappointing," revealing that "only a few regions showed increased activity across all the tasks, and they weren't very interesting brain regions." But the research team wasn't done. After failing to find regions that fired for many different activities, they decided to ask the opposite question: What, if anything, is active in the brain when someone is *not* trying to do a task? "It was an unusual question," notes Lieberman, but we should be glad they asked, because it led to a remarkable discovery: the team found that there's a particular set of regions in the brain that consistently activate when you're not attempting to do a cognitive task, and that just as consistently deactivate once you focus your attention on something specific.

Because almost any task caused this network to deactivate, the researchers originally called it "the task-induced deactivation network." Because this name was a mouthful, it was eventually abbreviated to a catchier label: "the default network."

At first, scientists had no idea what the default network did. They had a long list of tasks that turned it off (telling them what it didn't do), but little hard evidence about its true purpose. Even without good evidence, however, scientists began to develop intuitions based on their own experience. One of these pioneering thinkers is our guide to this research, Matthew Lieberman—who now enters our narrative as an active participant.

As Lieberman recalls, images of the default network were typically produced by asking a subject in a PET scanner to take a break from whatever repetitive activity the experiment required. Because the subject wasn't engaged in a specific task, it was easy for researchers to think of the default network as something that comes on when you're thinking about nothing. A little self-reflection, however, makes clear that our brains are hardly ever actually thinking about nothing. Even without a specific task, they tend to remain highly active, with thoughts and ideas flitting by in an ongoing noisy chatter. On further self-reflection, Lieberman realized that this background hum of activity tends to focus on a small number of targets: thoughts about "other people, yourself, or both." The default network, in other words, seems to be connected to *social cognition*.

Sure enough, once scientists knew what to look for, they discovered that the regions of the brain that defined the default network are "virtually identical" to the networks that light up during social cognition experiments. When given downtime, in other words, our brain defaults to thinking about our social life.

It's here that Lieberman's research takes an interesting twist. When he first encountered the conclusion that the default network is social, he wasn't impressed. Like others in his field, he noted that people naturally have a strong interest in their own social life, so it's not surprising that this is what they like to think about when bored. As Lieberman continued to study different aspects of social cognition, however, his opinion shifted. "I have since become convinced that I had the relationship between these networks backward," he writes. "And this reversal is tremendously important." He now believes "we are interested in the social world *because we are built to turn on the default network during our free time.*" Put another way, our brains adapted to automatically practice social thinking during any moments of cognitive downtime, and it's this practice that helps us become really interested in our social world.

Lieberman and his collaborators devised a clever series of experiments to confirm this hypothesis. In one study, for example, they found that the default network lights up during downtime even in newborns. The importance of finding this activity in infants is that they "clearly haven't cultivated an interest in the social world yet. . . . [The infants studied] cannot even focus their eyes." This behavior must therefore be instinctual.

In another study, researchers put (adult) subjects in a scanner and asked them to solve math problems. They discovered that when they gave the subjects a three-second break between problems—a duration too short for them to decide to start thinking about something else—the default network still fired up to fill the small gap, further indicating that this drive to think about social issues kicks in like a reflex.

This finding underscores the fundamental importance of social connections to human well-being. As Lieberman summarizes: "The brain did not evolve over millions of years to spend its free time practicing something irrelevant to our lives." But the default network is not the whole story. Additional studies by Lieberman and his collaborators uncovered other examples where evolution placed "big bets" on the importance of sociality by adapting other expensive systems to serve its needs.

The loss of social connection, for example, turns out to trigger the same system as physical pain—explaining why the death of a family member, a breakup, or even just a social snub can cause such distress. In one simple experiment, it was discovered that over-the-counter painkillers reduced social pain. Given the power of the pain system in driving our behavior, its connection to our social life underscores the importance of social relationships to our species' success.

Lieberman also discovered that the human brain devotes significant resources to two different major networks that work together toward the goal of *mentalizing*: helping us understand other people's minds, including how they are feeling and their intentions. Something as simple as a casual conversation with a store clerk requires massive amounts of neuronal computational power to take in and process a high-bandwidth stream of clues about what's going on in the clerk's mind. Though this "mind reading" feels natural to us, it's actually an amazingly complicated feat performed by networks honed over millions of years of evolution. It's exactly these highly adapted systems that are leveraged by the rock paper scissor champions who opened this chapter.

These experiments represent only some key highlights among many from a vast social cognitive neuroscience literature that all point to the same conclusion: humans are wired to be social. In other words, Aristotle was on the right track when he called us social animals, but it took the modern invention of advanced brain scanners to help us figure out how much he was likely understating this reality.

■ ■ ■

This highly adapted human interest in social connection is a fascinating piece of our evolutionary history. It's also, however, a reality that should concern any digital minimalist. The intricate brain networks described above evolved over millions of years in environments where interactions were always rich, face-to-face encounters, and social groups were small and tribal. The past two decades, by contrast, are characterized by the rapid spread of *digital communication tools*—my name for apps, services, or sites that enable people to interact through digital networks—which have pushed people's social networks to be much larger and much less local, while encouraging interactions through short, text-based messages and approval clicks that are orders of magnitude less information laden than what we have evolved to expect.

Perhaps predictably, this clash of old neural systems with modern innovations has caused problems. Much in the same way that the "innovation" of highly processed foods in the mid-twentieth century led to a global health crisis, the unintended side effects of digital communication tools—a sort of social fast food—are proving to be similarly worrisome.

THE SOCIAL MEDIA PARADOX

Determining the impact of digital communication tools on our psychological well-being is complicated. There's no shortage of scientific studies examining this topic, but different groups draw different conclusions from the same literature.

Consider two contrasting takes on this topic that were both published around the same time in 2017. The first was an NPR story posted in March of that year, which summarized the

results of a pair of big-deal new studies about the connection
between social media use and well-being. Both studies found
strong correlations between social media use and a range of
negative factors, from perceived isolation to poorer physical
health. The NPR story's title summarizes these findings well:
"Feeling Lonely? Too Much Time on Social Media Might
Be Why."

Not long after this NPR article came out, two members of
Facebook's internal research team published a blog post defending their service against a rising tide of criticism that had
begun in the aftermath of the contentious 2016 election. In
this post, the authors acknowledge that some uses of social
media might make people less happy, but then point to several
research studies that establish that "when used properly," these
services make subjects measurably happier. Using Facebook to
keep in touch with friends and loved ones, the authors note,
"brings us joy and strengthens our sense of community."

In other words, depending on whom you ask, social media
is either making us lonely or bringing us joy.

To better understand this general phenomenon of contrast-
ing conclusions, let's look closer at the specific studies sum-
marized above. One of the main positive articles cited by the
Facebook blog post was authored by Moira Burke, a data sci-
entist at the company (who also coauthored the blog post), and
Robert Kraut, a human computer interaction specialist at
Carnegie Mellon University. It was published in the *Journal of
Computer-Mediated Communication* in July 2016. In this study,
Burke and Kraut recruited a group of around 1,900 Facebook

users who agreed to quantify their current level of happiness when prompted. The researchers then used the Facebook server logs to connect specific social media activities with these well-being scores. They found that when users received "targeted" and "composed" information written by someone they know well (e.g., a comment sent by a family member), they felt better. On the other hand, receiving targeted and composed information from someone they didn't know well, or receiving a "like," or reading a status update broadcast to many people didn't correlate with improved well-being.

Another positive article cited in the Facebook post was authored by social psychologists Fenne Deters from the Freie Universität Berlin and Matthias Mehl from the University of Arizona. It appeared in a journal called *Social Psychology and Personality Science* back in September 2013. In this study, Mehl and Deters deployed a controlled experiment. During a one-week period, some subjects were asked to make more Facebook posts than normal, while the others were given no instructions. The experimental group who were asked to post more ended up reporting less loneliness than the control group during this week. Closer questioning revealed this was due primarily to feeling more connected to their friends on a daily basis.

These two studies seem to paint a compelling picture of social media boosting happiness and banishing loneliness. But let's now muddy the waters by considering the main two negative studies cited in the NPR article that came out during the same period as the Facebook post.

The first of these studies was authored by a large team from diverse disciplines, led by Brian Primack from the University of Pittsburgh. It appeared in the prestigious *American Journal of Preventive Medicine* in July 2017. Primack and his team surveyed a nationally representative sample of adults between the ages of nineteen and twenty-two, using the same type of random sample techniques that pollsters deploy to measure public opinion during elections. The survey asked a standard set of questions that measure the subject's perceived social isolation (PSI)—a loneliness metric. It also asked about usage of eleven different major social media platforms. After crunching the numbers, the researchers found that the more someone used social media, the more likely they were to be lonely. Indeed, someone in the highest quartile of social media use was three times more likely to be lonelier than someone in the lowest quartile. These results held up even after the researchers controlled for factors such as age, gender, relationship status, household income, and education. Primack admitted to NPR that he was surprised by the results: "It's social media, so aren't people supposed to be socially connected?" But the data was clear. The more time you spend "connecting" on these services, the more isolated you're likely to become.

The other study cited in the NPR article was authored by Holly Shakya of the University of California–San Diego and Nicholas Christakis of Yale, and it appeared in the *American Journal of Epidemiology* in February 2017. Shakya and Christakis used data from over 5,200 subjects from a nationally representative panel survey, combined with observed Facebook behavior of the subjects. They studied associations between Facebook activity and self-reported measures of physical health, mental health, and life satisfaction (among other quality of life metrics). As they report: "Our results show that overall, the use of Facebook was negatively associated with well-being." They found, for example, that if you increase the amount of likes or links clicked by a standard deviation, mental health decreases by 5 to 8 percent of a standard deviation. These negative connections still held when, like in the Primack study, they controlled for relevant demographic variables.

These dueling studies seem to present a paradox—social media makes you feel both connected and lonely, happy and sad. To resolve this paradox, let's start by looking closer at the experimental designs described above. The studies that found positive results focused on *specific behaviors* of social media users, while the studies that found negative results focused on *overall use* of these services. The natural assumption is that these variables would be positively connected: If common social media behaviors increase well-being, then the more you use these services, the more of these mood-boosting behaviors you'll engage in, and the happier you should be. Therefore, after reading the positive studies, you would expect that increasing social media use would increase well-being—but this, of course, was the opposite of what the researchers discovered in the negative studies.

There must, therefore, be another factor at play—something that increases the more you use social media, generating negative impacts that swamp out the smaller positive boosts. Fortunately for our investigation, Holly Shakya identified a likely suspect for this factor: the more you use social media to interact with your network, the less time you devote to offline communication. "What we know at this point," Shakya told NPR, "is that we have evidence that replacing your real-world relationships with social media use is detrimental to your well-being."

To help explore this idea, Shakya and Christakis also measured offline interactions and found they were associated with positive effects—a finding that has been widely replicated in the social psychology literature. As they then noted, the negative associations of Facebook use are comparable in magnitude to the positive impact of offline interaction—suggesting a trade-off.

The problem, then, is not that using social media directly makes us unhappy. Indeed, as the positive studies cited above found, certain social media activities, when isolated in an experiment, modestly boost well-being. The key issue is that using social media tends to take people away from the real-world socializing that's massively more valuable. As the negative studies imply, the more you use social media, the less time you tend to devote to offline interaction, and therefore the worse this value deficit becomes—leaving the heaviest social media users much more likely to be lonely and miserable. The small boosts you receive from posting on a friend's wall or liking their latest Instagram photo can't come close to compensating for the large loss experienced by no longer spending real-world time with that same friend.

As Shakya summarizes: "Where we want to be cautious . . . is when the sound of a voice or a cup of coffee with a friend is replaced with 'likes' on a post."

■ ■ ■

The idea that real-world interactions are more valuable than online interactions isn't surprising. Our brains evolved during a period when the *only* communication was offline and face-to-face. As argued earlier in the chapter, these offline interactions are incredibly rich because they require our brains to process large amounts of information about subtle analog cues such as body language, facial expressions, and voice tone. The low-bandwidth chatter supported by many digital communication tools might offer a simulacrum of this connection, but it leaves most of our high-performance social processing networks underused—reducing these tools' ability to satisfy our intense sociality. This is why the value generated by a Facebook comment or Instagram like—although real—is minor compared to the value generated by an analog conversation or shared real-world activity.

We don't have good data on why people trade online for offline communication when given access to digital communication tools, but it's easy to generate convincing hypotheses based on common experience. An obvious culprit is that online interaction is both easier and faster than old-fashioned conversation. Humans are naturally biased toward activities that require less energy in the short term, even if it's more harmful in the long term—so we end up texting our sibling instead of calling them on the phone, or liking a picture of a friend's new baby instead of stopping by to visit.

A subtler effect is the way that digital communication tools can subvert the offline communication that remains in your life. Because our primal instinct to connect is so strong, it's difficult to resist checking a device in the middle of a conversation with a friend or bath time with a child—reducing the quality of the richer interaction right in front of us. Our analog brain cannot easily distinguish between the importance of the person in the room with us and the person who just sent us a new text.

Finally, as detailed in the first part of this book, many of these tools are engineered to hijack our social instincts to create an addictive allure. When you spend multiple hours a day compulsively clicking and swiping, there's much less free time left for slower interactions. And because this compulsive use emits a patina of socialness, it can delude you into thinking that you're already serving your relationships well, making further action unnecessary.

To state the obvious, this account doesn't cover all the possible dangers of digital communication tools. Critics have also highlighted the ability for social media to make us feel ostracized or inadequate, as well as to stoke exhausting outrage, inflame our worst tribal instincts, and perhaps even degrade the democratic process itself. For the remainder of this chapter, however, I want to bypass a discussion of the potential pathologies of the social media universe and keep our focus on the zero-sum relationship between online and offline interaction. I believe this to be the most fundamental of the issues caused by the digital communication era, and the key trap that a minimalist must understand in trying to successfully navigate the pluses and minuses of these new tools.

RECLAIMING CONVERSATION

Up to this point in the chapter, we've relied on some clunky terminology to differentiate interaction mediated through text interfaces and mobile screens from the old-fashioned analog communication our species evolved to crave. Going forward, I want to borrow some useful phrasing from MIT professor Sherry Turkle, a leading researcher on the subjective experience of technology. In her 2015 book, *Reclaiming Conversation*, Turkle draws a distinction between *connection*, her word for the low-bandwidth interactions that define our online social lives, and *conversation*, the much richer, high-bandwidth communication that defines real-world encounters between humans. Turkle agrees with our premise that conversation is crucial:

> Face-to-face conversation is the most human—and humanizing—thing we do. Fully present to one another, we learn to listen. It's where we develop the capacity for empathy. It's where we experience the joy of being heard, of being understood.

In her book, Turkle presents anthropological case studies that highlight the same "flight from conversation" that was captured by the quantitative studies cited earlier in this chapter, and in doing so, she puts a human face on the decreased well-being that occurs when conversation is replaced with connection.

Turkle, for example, introduces her readers to middle school students who struggle with empathy, as they lack the practice of reading facial cues that comes from conversation, as well as a thirty-four-year-old colleague who comes to realize her online interactions all have an exhausting element of performance that have led her to the point where the line between real and performed is blurring. Turning her attention to the workplace, Turkle finds young employees who retreat to email because the thought of an unstructured conversation terrifies them, and unnecessary office tensions that fester when communication shifts from nuanced conversation to ambiguous connection.

During an appearance on *The Colbert Report*, host Stephen Colbert asked Turkle a "profound" question that gets at the core of her argument: "Don't all these little tweets, these little sips of online connection, add up to one big gulp of real conversation?" Turkle was clear in her answer: *No, they do not.* As she expands: "Face-to-face conversation unfolds slowly. It teaches patience. We attend to tone and nuance." On the other hand: "When we communicate on our digital devices, we learn different habits."

As a true digital minimalist, Turkle approaches these issues from a standpoint of smarter use of digital communication tools, not blanket abstention. "My argument is not anti-technology," she writes. "It's pro-conversation." She's confident that we can make the necessary changes to reclaim the conversation we need to thrive, noting that despite the "seriousness of the moment" she remains optimistic that once we recognize the issues in replacing conversation with connection, we can rethink our practices.

I share Turkle's optimism that there's a minimalist solution to this problem, but I'm more pessimistic about the magnitude of effort required. Toward the end of her book, Turkle offers a series of recommendations, which center in large part on the idea of making more space in your life for quality conversation. The objective of this recommendation is faultless, but its effectiveness is questionable. As argued earlier in this chapter, digital communication tools, if used without intention, have a way of forcing a trade-off between conversation and connection. If you don't first reform your relationship with tools like social media and text messaging, attempts to shoehorn more conversation into your life are likely to fail. It can't simply be digital business as usual *augmented* with more time for authentic conversation—the shift in behavior will need to be more fundamental.

To succeed with digital minimalism, you have to confront this rebalancing between conversation and connection in a way that makes sense to you. To prime your thinking along these lines, however, I'll present in the following pages a somewhat radical solution—a philosophy of sorts for socializing in a digital age—that I personally find to be appealing. I refer to this philosophy by the superfluously alliterative name *conversation-centric communication.* You can moderate these ideas as needed to accommodate the idiosyncratic realities of your social life, or reject them altogether—but you cannot avoid the need to think about solutions to these issues that are comparably aggressive.

■ ■ ■

Many people think about conversation and connection as two different strategies for accomplishing the same goal of maintaining their social life. This mind-set believes that there are many different ways to tend important relationships in your life, and in our current modern moment, you should use all tools available—spanning from old-fashioned face-to-face talking, to tapping the heart icon on a friend's Instagram post.

The philosophy of conversation-centric communication takes a harder stance. It argues that conversation is the only form of interaction that in some sense *counts* toward maintaining a relationship. This conversation can take the form of a face-to-face meeting, or it can be a video chat or a phone call—so long as it matches Sherry Turkle's criteria of involving nuanced analog cues, such as the tone of your voice or facial expressions. Anything textual or non-interactive—basically, all social media, email, text, and instant messaging—doesn't count as conversation and should instead be categorized as mere connection.

In this philosophy, connection is downgraded to a logistical role. This form of interaction now has two goals: to help set up and arrange conversation, or to efficiently transfer practical information (e.g., a meeting location or time for an upcoming event). Connection is no longer an alternative to conversation; it's instead its supporter.

If you subscribe to conversation-centric communication, you might still maintain some social media accounts for the purposes of logistical expediency, but gone will be the habit of regularly browsing these services throughout your day, sprinkling "likes" and short comments, or posting your own updates and desperately checking for the feedback they accrue. With this in mind, there would no longer be much purpose in keeping these apps on your phone, where they will mainly serve to undermine your attempts at richer interaction. They would instead more productively reside on your computer, where they're occasionally put to specific use.

Similarly, if you adopt conversation-centric communication, you'll still likely rely on text-messaging services to simplify information gathering, or to coordinate social events, or to ask quick questions, but you'll no longer participate in open-ended, ongoing text-based conversations throughout your day. The socializing that counts is real conversation, and text is no longer a sufficient alternative.

Notice, in true minimalist fashion, conversation-centric communication doesn't ask that you abandon the wonders of digital communication tools. On the contrary, this philosophy recognizes that these tools can enable significant improvements to your social life. Among other advantages, these new technologies greatly simplify the process of arranging conversation. When you unexpectedly find yourself free on a weekend afternoon, a quick round of text messages can efficiently identify a friend available to join you for a walk. Similarly, a social media service might alert you that an old friend is going to be in town, prompting you to arrange a dinner.

Innovations in digital communication also provide cheap and effective ways to banish the obstacle of distance in seeking conversation. When my sister was living in Japan, we would regularly converse over FaceTime, deciding to place a call based on the same spur-of-the-moment inspiration with which you might casually drop in on a relative living down the street. At any other period of human history, this capability would be considered miraculous. In short, this philosophy has nothing against technology—so long as the tools are put to use to improve your real-world social life as opposed to diminishing it.

To be clear, conversation-centric communication requires sacrifices. If you adopt this philosophy, you'll almost certainly reduce the number of people with whom you have an active relationship. Real conversation takes time, and the total number of people for which you can uphold this standard will be significantly less than the total number of people you can follow, retweet, "like," and occasionally leave a comment for on social media, or ping with the occasional text. Once you no longer count the latter activities as meaningful interaction, your social circle will seem at first to contract.

This sense of contraction, however, is illusory. As I have

argued throughout this chapter, conversation is the good stuff; it's what we crave as humans and what provides us with the sense of community and belonging necessary to thrive. Connection, on the other hand, though appealing in the moment, provides very little of what we need.

In the early days of adopting a conversation-centric mindset, you might miss the security blanket of what Stephen Colbert astutely labeled "little sips of online connection," and the sudden loss of weak ties to the fringes of your social network might induce moments of loneliness. But as you trade more of this time for conversation, the richness of these analog interactions will far outweigh what you're leaving behind. In her book, Sherry Turkle summarizes research that found just five days at a camp with no phones or internet was enough to induce major increases in the campers' well-being and sense of connection. It won't take many walks with a friend, or pleasantly meandering phone calls, before you begin to wonder why you previously felt it was so important to turn away from the person sitting right in front of you to leave a comment on your cousin's friend's Instagram feed.

■ ■ ■

Whether or not you accept my proposed philosophy of conversation-centric communication, I hope you do accept its motivating premise: the relationship between our deeply human sociality and modern digital communication tools is fraught and can produce significant issues in your life if not handled carefully. You cannot expect an app dreamed up in a dorm room, or among the Ping-Pong tables of a Silicon Valley incubator, to successfully replace the types of rich interactions to which we've painstakingly adapted over millennia. Our sociality is simply too complex to be outsourced to a social network or reduced to instant messages and emojis.

Any digital minimalist must confront this reality and manage his or her relationship with these tools accordingly. I'm an advocate for deploying a conversation-centric approach for this purpose, because I fear any attempt to maintain a two-tier approach to conversation—combining digital communication with old-fashioned analog conversation—will ultimately falter. That being said, others might be stronger than I am when it comes to maintaining a healthy balance between these two interactive magisterium, so I'll resist the urge for dogmatism on this point. The key is the intention behind what you decide, not necessarily its details.

To aid this minimalist pondering, this chapter ends with a collection of concrete practices to help you reclaim conversation. My now standard caveats apply: These suggestions are neither comprehensive nor obligatory. They instead provide you with a sense of the *types* of decisions you can make to help move back toward the type of communication we're adapted to crave.

PRACTICE: DON'T CLICK "LIKE"

Contrary to popular lore, Facebook didn't invent the "Like" button. That credit goes to the largely forgotten FriendFeed service, which introduced this feature in October 2007. But when the massively more popular Facebook introduced the iconic thumbs-up icon sixteen months later, the trajectory of social media was forever changed.

The initial announcement of the feature, posted by a corporate communications officer named Kathy Chan in the winter of 2009, reveals a modest motivation for the innovation. As Chan explains, many Facebook posts attracted a large number of comments that were all saying more or less the same thing; e.g., "Great!" or "I love it!" The "Like" button was introduced as a simpler way to indicate your general approval of a post, which would both save time and allow the comments to be reserved for more interesting notes.

As I explored in the first part of this book, from these humble beginnings, the "Like" feature evolved to become the foundation on which Facebook rebuilt itself from a fun amusement that people occasionally checked, to a digital slot machine that began to dominate its users' time and attention. This button introduced a rich new stream of social approval indicators that arrive in an unpredictable fashion—creating an almost impossibly appealing impulse to keep checking your account. It also provided Facebook much more detailed information on your preferences, allowing their machine-learning algorithms to digest your humanity into statistical slivers that could then be mined to push you toward targeted ads and stickier content. Not surprisingly, almost every other successful major social media platform soon followed FriendFeed and Facebook's lead and added similar one-click approval features to their services.

In the context of this chapter, however, I don't want to focus on the boon the "Like" button proved to be for social media companies. I want to instead focus on the harm it inflicted to our human need for real conversation. To click "Like," within the precise definitions of information theory, is literally the least informative type of nontrivial communication, providing only a minimal *one bit* of information about the state of the sender (the person clicking the icon on a post) to the receiver (the person who published the post).

Earlier, I cited extensive research that supports the claim that the human brain has evolved to process the flood of information generated by face-to-face interactions. To replace this rich flow with a single bit is the ultimate insult to our social processing machinery. To say it's like driving a Ferrari under the speed limit is an understatement; the better simile is towing a Ferrari behind a mule.

■ ■ ■

Motivated by the above observations, this practice suggests that you transform the way you think about the different flavors of one-click approval indicators that populate the social media universe. Instead of seeing these easy clicks as a fun way to nudge a friend, start treating them as poison to your attempts to cultivate a meaningful social life. Put simply, you should stop using them. Don't click "Like." Ever. And while you're at it, stop leaving comments on social media posts as well. No "so cute!" or "so cool!" Remain silent.

The reason I'm suggesting such a hard stance against these seemingly innocuous interactions is that they teach your mind that connection is a reasonable alternative to conversation. The motivating premise behind my conversation-centric communication philosophy is that once you accept this equality, despite your good intentions, the role of low-value interactions will inevitably expand until it begins to push out the high-value socializing that actually matters. If you eliminate these trivial interactions cold turkey, you send your mind a clear message: conversation is what counts—don't be distracted from this reality by the shiny stuff on your screen. As I mentioned before, you may think you can balance both types of interaction, but most people can't.

Some worry that this sudden abstention from social media nudges will annoy people in their social circle. One person I mentioned this strategy to, for example, expressed concern that if she didn't leave a comment on a friend's latest baby picture, it would be noted as a callous omission. If the friendship is important, however, let the concern about this reaction motivate you to invest the time required to set up a real conversation. Actually visiting the new mom will return significantly more value to both of you than adding a short "awww!" to a perfunctory scroll of comments.

If you couple this push toward more conversation with a blanket warning to your circle that you're "not using social media much these days," you'll effectively insulate yourself from most complaints this policy might create. The person cited above, for example, ended up bringing a meal to her new-mother friend. This one act strengthened the relationship and increased well-being more than a hundred quick social media reactions could have.

Finally, it's worth noting that refusing to use social media icons and comments to interact means that some people *will* inevitably fall out of your social orbit—in particular, those whose relationship with you exists only over social media. Here's my tough love reassurance: let them go. The idea that it's valuable to maintain vast numbers of weak-tie social connections is largely an invention of the past decade or so—the detritus of overexuberant network scientists spilling inappropriately into the social sphere. Humans have maintained rich and fulfilling social lives for our entire history without needing the ability to send a few bits of information each month to people we knew briefly during high school. Nothing about your life will notably diminish when you return to this steady state. As an academic who studies and teaches social media explained to me: "I don't think we're meant to keep in touch with so many people."

To summarize, the question of whether or not you continue to use social media as a digital minimalist, and on what terms, is complicated and depends on many different factors. But regardless of what final decisions you make along these lines, I urge you, for the sake of your social well-being, to adopt the baseline rule that you'll no longer use social media as a tool for low-quality relationship nudges. Put simply, don't click and don't comment. This basic stricture will radically change for the better how you maintain your social life.

PRACTICE: CONSOLIDATE TEXTING

A major obstacle in attempting to shift your social life from connection back to conversation is the degree to which text communication—be it delivered through SMS, iMessage, Facebook Messenger, or WhatsApp—now pervades the very definition of friendship. Sherry Turkle, who has been studying phone use since the beginning of the smartphone era, describes this reality as follows:

> Phones have become woven into a fraught sense of obligation in friendship. . . . Being a friend means being "on call"—tethered to your phone, ready to be attentive, online.

In the last practice, I recommended that you stop interacting with friends through social media "likes" and comments. This might raise some eyebrows, but with enough apologetic shrugging, and a commitment to replace these low-value clicks with higher-value conversation, the change will be accepted. For many people, however, leaving the world of text messaging would prove substantially more disruptive. Friendship doesn't require Facebook "likes," but if you're below a certain age, it does seem to require texting. To shirk your duty to be "on call" in this way would be a serious abdication.

This state of affairs presents a quandary. Earlier in this chapter, I argued that text messaging is not sufficiently rich to fulfill our brain's craving for real conversation. The more you text, however, the less necessary you'll deem real conversation, and, perversely, when you do interact face-to-face, your compulsion to keep checking on other interactions on your phone will diminish the value you experience. We're left, then, with a technology that's required in your social life while simultaneously reducing the value you derive from it. As someone who is keenly aware of these tensions, I want to offer a compromise that respects both your obligation to be "on call" and your human craving for real conversation: consolidate texting.

■ ■ ■

This practice suggests that you keep your phone in Do Not Disturb mode by default. On both iPhones and Android devices, for example, this mode turns off notifications when text messages arrive. If you're worried about emergencies, you can easily adjust the settings so calls from a selected list (your spouse, your kid's school) do come through. You can also set a schedule that turns the phone to this mode automatically during predetermined times.

When you're in this mode, text messages become like emails: if you want to see if anyone has sent you something, you must turn on your phone and open the app. You can now schedule specific times for texting—consolidated sessions in which you go through the backlog of texts you received since the last check, sending responses as needed and perhaps even having some *brief* back-and-forth interaction before apologizing that you have to go, turning the phone back to Do Not Disturb mode, and continuing with your day.

There are two major motivations for this practice. The first is that it allows you to be more present when you're not texting. Once you no longer treat text interactions as an ongoing conversation that you must continually tend, it's much easier to concentrate fully on the activity before you. This will increase the value you get out of these real-world interactions. It might also provide some anxiety reduction, as our brains don't react well to constant disruptive interaction (see the previous chapter on the importance of solitude).

The second motivation for this practice is that it can upgrade the nature of your relationships. When your friends and family are able to instigate meandering pseudo-conversations with you over text at any time, it's easy for them to become complacent about your relationship. These interactions give the appearance of close connection (even though, in reality, they're far from this standard), providing a disincentive to invest more time in more meaningful engagement.

On the other hand, if you only check your text messages occasionally, this dynamic changes. They're still able to send you questions and get back a response in a reasonable amount of time, or send you a reminder and be sure that you'll see it. But these more asynchronous and logistical interactions no longer give off the approximate luster of true conversation. The result is that both of you will be more motivated to fill this void with better interaction, as the relationship will seem strained in the absence of back-and-forth dialogue.

Being less available over text, in other words, has a way of paradoxically strengthening your relationship even while making you (slightly) less available to those you care about. This point is crucial because many people fear that their relationships will suffer if they downgrade this form of lightweight connection. I want to reassure you that it will instead strengthen the relationships you care most about. You can be the one person in their life who actually *talks* to them on a regular basis, forming a deeper, more nuanced relationship than any number of exclamation points and bitmapped emojis can provide.

This all being said, the practice of consolidating texting might still cause trouble. If people are used to grabbing your attention at any time, then your new absence will cause occasional consternation. But these concerns are easy to resolve. Simply tell people close to you that you check texts several times a day, so if they send you something, you'll see it shortly, and that if they need you urgently, they can always call you (it's here that you should configure your Do Not Disturb mode settings to let in calls from a favored list). This response calms any legitimate concerns about your availability while still freeing you from an unrelenting duty to your messages.

To conclude, let's agree on the obvious claim that text messaging is a wonderful innovation that makes many parts of life significantly more convenient. This technology only becomes a problem when you treat it as a reasonable alternative to real conversation. By simply keeping your phone in Do Not Disturb mode by default, and making texts something you check on a regular schedule—not a persistent background source of ongoing chatter—you can maintain the major advantages of the technology while sidestepping its more pernicious effects.

PRACTICE: HOLD CONVERSATION OFFICE HOURS

For over a century, the telephone has provided a way to engage in high-quality conversation over long distances. This remarkable achievement helped satisfy social cravings in an age where we no longer spent our whole lives in tight-knit tribes. The problem with phones, of course, is the inconvenience of *placing* calls. Without being able to see the person you're about to interrupt with a request to chat, you have no way of knowing whether or not your interaction will be well received. I still vividly remember my childhood anxiety when placing calls to friends—not knowing who from their family would pick up and how they would feel about the intrusion. With this shortcoming in mind, we should perhaps not be surprised that as soon as easier communication technologies were introduced—text messages, emails—people seemed eager to abandon this time-tested method of conversation for lower-quality connections (Sherry Turkle calls this effect "phone phobia").

Fortunately, there's a simple practice that can help you side-step these inconveniences and make it much easier to regularly enjoy rich phone conversations. I learned it from a technology executive in Silicon Valley who innovated a novel strategy for supporting high-quality interaction with friends and family: he tells them that he's always available to talk on the phone at 5:30 p.m. on weekdays. There's no need to schedule a conversation or let him know when you plan to call—just dial him up. As it turns out, 5:30 is when he begins his traffic-clogged commute home in the Bay Area. He decided at some point that he wanted to put this daily period of car confinement to good use, so he invented the 5:30 rule.

The logistical simplicity of this system enables this executive to easily shift time-consuming, low-quality connections into higher-quality conversation. If you write him with a somewhat complicated question, he can reply, "I'd love to get into that. Call me at 5:30 any day you want." Similarly, when I was visiting San Francisco a few years back and wanted to arrange a get-together, he replied that I could catch him on the phone any day at 5:30, and we could work out a plan. When he wants to catch up with someone he hasn't spoken to in a while, he can send them a quick note saying, "I'd love to get up to speed on what's going on in your life, call me at 5:30 sometime." His close friends and family members, I assume, have long since internalized the 5:30 rule, and probably feel more comfortable calling him on a whim than they do other people in their circles, as they know he's available then and always happy to take their call.

This executive enjoys a more satisfying social life than most people I know, even though he works in demanding technology start-ups that take up a lot of his time. He hacked his schedule in such a way that eliminated most of the overhead related to conversation and therefore allowed him to easily serve his human need for rich interaction. Perhaps not surprisingly, I want to propose here that you follow his lead.

■ ■ ■

This practice suggests that you follow the aforementioned executive's example by instating your own variation of his *conversation office hours* strategy. Put aside set times on set days during which you're always available for conversation. Depending on where you are during this period, these conversations might be exclusively on the phone or could also include in-person meetings. Once these office hours are set, promote them to the people you care about. When someone instigates a low-quality connection (say, a text message conversation or social media ping), suggest they call or meet you during your office hours sometime when it is convenient for them. Similarly, once office hours are in place, it's easy to reach out proactively to people you care about and invite them to converse with you during these hours whenever they're next available.

I've seen several variations of this practice work well. Using a commute for phone conversations, like the executive introduced above, is a good idea if you follow a regular commuting schedule. It also transforms a potentially wasted part of your day into something meaningful. Coffee shop hours are also popular. In this variation, you pick some time each week during which you settle into a table at your favorite coffee shop with the newspaper or a good book. The reading, however, is just the backup activity. You spread the word among people you know that you're always at the shop during these hours with the hope that you soon cultivate a rotating group of regulars that come hang out. I first witnessed this strategy in a coffee shop in a town near where I grew up. There's a small group of late-middle-aged men who set up shop on Saturday mornings and pull friends into their conversational orbit as they stop in the shop throughout the day. Taking a page out of the English cultural playbook, you can also consider running these office hours once a week during happy hour at a favored bar.

I've also seen people deploy daily walks for this purpose. Steve Jobs was famous for his long strolls around the tree-lined Silicon Valley neighborhood where he lived. If you were in his inner circle, you could expect invitations to join him for what was sure to be an intense conversation. Ironically for the inventor of the iPhone, Jobs was not the type of person who would be interested in maintaining important relationships through ongoing drips of digital pings.

In my own life as a professor, I transformed my actual office hours into something broader. In my field, you're required to put aside some time once a week for students in your classes to stop by to ask questions. Early in my career at Georgetown, I realized these sessions held value well beyond just interacting with my current students. I now try to expand the length of my office hours so I can declare them open to all Georgetown students. When any student writes me to ask a question, or request advice, or share their experience with one of my books, I can point them to my regular office hours and say, "Stop by or call anytime." And they do. The result is that I'm much better connected to the student body at my university than I would be if I were still trying to arrange a custom-scheduled interaction for every request that came my way.

The conversation office hours strategy is effective for improving your social life because it overcomes the major obstacle to meaningful socializing: the concern, mentioned above, that unsolicited calls might be bothersome. People crave real conversation, but this obstacle is often enough to prevent it. If you remove it by holding conversation office hours, you'll be surprised by how many more of these rewarding interactions you can now fit into your normal week.

6

Reclaim Leisure

LEISURE AND THE GOOD LIFE

In his *Nicomachean Ethics*, compiled in the fourth century BC, Aristotle tackles a question as urgent then as it is today: How does one live a good life? The *Ethics* divides its answer across ten books. Much of the first nine focus on what Aristotle calls "practical virtues," such as fulfilling your duties, or acting justly when faced with injustice and courageously when faced with danger. But then, in the tenth and final book of the *Ethics*, Aristotle steps back from this gritted-teeth heroic virtue and makes a radical turn in his argument: "The best and most pleasant life is the life of the intellect." He concludes, "This life will also be the happiest."

As Aristotle elaborates, a life filled with deep thinking is happy because contemplation is an "activity that is appreciated for its own sake . . . nothing is gained from it except the act of contemplation." In this offhand claim, Aristotle is identifying, for perhaps the first time in the history of recorded philosophy, an idea that has persisted throughout the intervening millennia and continues to resonate with our understanding of human nature today: a life well lived requires activities that serve no other purpose than the satisfaction that the activity itself generates.

As the MIT philosopher Kieran Setiya expands in his modern interpretation of the *Ethics*, if your life consists only of actions whose "worth depends on the existence of problems, difficulties, needs, which these activities aim to solve," you're vulnerable to the existential despair that blooms in response to the inevitable question, *Is this all there is to life?* One solution to this despair, he notes, is to follow Aristotle's lead and embrace pursuits that provide you a "source of inward joy."

In this chapter, I call these joyful activities *high-quality leisure*. The reason that I'm reminding you here of their importance to a well-crafted life—an idea that dates back over two thousand years—is that I've become convinced that to successfully tame the problems of our modern digital world, you must both understand and deploy the core insights of this ancient wisdom.

■ ■ ■

To explain my claimed connection between high-quality leisure and digital minimalism, it's useful to first highlight a related phenomenon. Those of us who study the intersection of technology and culture are well read in the small but popular journalistic subgenre in which the author describes the experience of taking a temporary break from modern technologies. These intrepid souls almost always report that the disconnection generates a feeling of emotional distress. Here, for example, is the social critic Michael Harris describing his experience spending a week without the internet or cell service in a rustic cabin:

> By the end of day two . . . I miss everyone. I miss my bed and my television and Kenny and dear old Google. I stare hopelessly for an hour at the ocean, a coruscating kind of liquid metal; I feel the urge to change the channel every ten minutes. But the same water goes on and on, like a decree. Torture.

This distress is often explained in the terminology of addiction, in which it can be cast as withdrawal symptoms experienced by an addict. ("I remember that this was never going to be easy, that withdrawal symptoms are to be expected," writes Harris about his experience at the cabin.) But this interpretation is problematic. As we explored in part 1 of this book, the psychological forces that lead us to compulsively use technology are typically best understood as moderate behavioral addictions—which can make technology very alluring when it's around, but aren't nearly as severe as chemical dependency. This explains why this distress is often described as more diffuse and abstract than the strong and specific cravings felt by a substance addict going through classic withdrawal.

It's not that Harris had a specific online activity that he

really missed (like a smoker without his cigarettes), it's instead that he was uncomfortable about not having access in general. This distinction is subtle, but it's also crucial for understanding the productive connection between Aristotle and digital minimalism. The more I study this topic, the more it becomes clear to me that low-quality digital distractions play a more important role in people's lives than they imagine. In recent years, as the boundary between work and life blends, jobs become more demanding, and community traditions degrade, more and more people are failing to cultivate the high-quality leisure lives that Aristotle identifies as crucial for human happiness. This leaves a void that would be near unbearable if confronted, *but* that can be ignored with the help of digital noise. It's now easy to fill the gaps between work and caring for your family and sleep by pulling out a smartphone or tablet, and numbing yourself with mindless swiping and tapping. Erecting barriers against the existential is not new—before YouTube we had (and still have) mindless television and heavy drinking to help avoid deeper questions—but the advanced technologies of the twenty-first-century attention economy are particularly effective at this task.

Harris felt uncomfortable, in other words, not because he was craving a particular digital habit, but because he didn't know what to do with himself once his general access to the world of connected screens was removed.

If you want to succeed with digital minimalism, you cannot ignore this reality. If you begin decluttering the low-value digital distractions from your life *before* you've convincingly filled in the void they were helping you ignore, the experience will be unnecessarily unpleasant at best and a massive failure at worse. The most successful digital minimalists, therefore, tend to start their conversion by renovating what they do with their free time—cultivating high-quality leisure before culling the worst of their digital habits. In fact, many minimalists will describe a phenomenon in which digital habits that they previously felt to be essential to their daily schedule suddenly seemed frivolous once they became more intentional about what they did with their time. When the void is filled, you no longer need distractions to help you avoid it.

Inspired by these observations, the goal of this chapter is to help you cultivate high-quality leisure in your own life. The three sections that follow each explore a different lesson about what properties define the most rewarding leisure activities. These are followed by a discussion of the somewhat paradoxical role new technology plays in these activities, and then a collection of concrete practices that can help you get started cultivating these high-quality pursuits.

THE BENNETT PRINCIPLE

A useful place to start investigating high-quality leisure is within the so-called FI community. For those who are unfamiliar with this trend, the acronym *FI* stands for *financial independence*, which refers to the pecuniary state in which your assets produce enough income to cover your living expenses.

Many people think of FI as a goal you reach around retirement age, or perhaps after receiving a large inheritance, but in recent years the internet helped fuel a newly resurgent FI community that consists mainly of young people who are finding shortcuts to this freedom through extreme frugality.

Most of the attention on the FI 2.0 movement focuses on its underlying financial insights,* but these details are not relevant for our purposes. What does matter is the fact that these financially independent young people provide particularly good case studies for exploring high-quality leisure. There are two reasons for this claim. First, and perhaps most obvious, when you achieve FI, you suddenly have many more leisure hours to fill than the average person. The second reason is that the subversive decision to pursue FI at a young age, which typically leads to radical lifestyle decisions, self-selects for individuals who are unusually intentional about how they live

---

* For those who *are* interested, the central insight of the FI 2.0 movement is that if you can radically reduce your living expenses, you gain two advantages: (1) you can save money at a much faster pace (a 50 to 70 percent savings rate is common), and (2) you don't have to save as much to become independent, as the expenses you need to meet are lower. If you need only $30,000 take-home pay to live comfortably, for example, then saving $750,000 in a low-cost index fund will likely cover these expenses (with inflation adjustments) for decades. Now imagine that you're a young couple with two good salaries that generate $100,000 in take-home pay each year. Because you need only $30,000 to live on, you can save $70,000 a year. Assuming a 5 to 6 percent annual growth rate, you'd hit your target in eight to nine years. If you start this in your twenties, you'll end up financially independent by your late thirties. Naturally, much of the FI 2.0 literature focuses on the argument that these levels of frugality are less drastic than you might imagine.

their lives. This combination of abundant free time and commitment to intentional living makes this group an ideal source of insight into effective leisure.

Let's start this search for insight by interrogating the habits of the informal leader of the FI 2.0 movement: a former engineer named Pete Adeney, who became financially independent in his early thirties and now blogs about his life under the purposefully self-deprecating moniker Mr. Money Mustache. When Pete became financially independent, he didn't fill his life with the types of passive leisure activities we often associate with young men relaxing—playing video games, watching sports, web surfing, long evenings at the bar—he instead leveraged his freedom to become even more active.

Pete doesn't own a television and doesn't subscribe to Netflix or Hulu. He occasionally rents a movie on Google Play, but for the most part, his family doesn't use screens to provide entertainment. Where he does spend most of his time is working on projects. Preferably outside. Here's how Pete explains his leisure philosophy on his blog:

> I never understood the joy of watching other people play sports, can't stand tourist attractions, don't sit on the beach unless there's a really big sand castle that needs to be made, [and I] don't care about what the celebrities and politicians are doing. . . . Instead of all this, I seem to get satisfaction only from making stuff. Or maybe a better description would be solving problems and making improvements.

In recent years, Pete renovated his family's home and then built a standalone outbuilding in their yard to serve as an office and music studio. These projects completed, and eager for more holes to dig and drywall to hang, he somewhat impulsively bought a run-down retail building on the main street of his hometown of Longmont, Colorado. He's currently transforming it into what he calls Mr. Money Mustache World Headquarters. What, exactly, he plans to do with the space once finished isn't yet quite clear—but the end goal isn't really the point; he seems to have invested in this building in large part for the project. As Pete summarizes his leisure philosophy: "If you leave me alone for a day . . . I'll have a joyful time rotating between carpentry, weight training, writing, playing around with instruments in the music studio, making lists and executing tasks from them."

We can find a similar commitment to action in the lifestyle of Liz Thames, who also reached financial independence in her early thirties and blogs about it on the popular Frugalwoods website. Upon achieving FI, Liz and her husband, Nate, pushed their enjoyment of activity to a new extreme—leaving their home in bustling Cambridge, Massachusetts, and moving onto a sixty-six-acre homestead sited on the side of a small mountain in rural Vermont.

As Liz explained to me when I asked her about this decision, moving to a homestead of this size was not a choice made lightly. Their long gravel driveway, for example, requires constant maintenance. If a tree falls, it needs to be sawed and removed, "even if it's ten below outside." If it's snowing, they must plow often, or the snow pile will become too deep for the tractor to push, trapping them on their property—which is not ideal, as their nearest neighbor is a long hike away and they don't have cell service to let them know they need help.

Liz and Nate heat their home with wood from their property, which also turns out to require quite a bit of effort. "We spend the whole summer harvesting wood," Liz told me. "You have to go into the forest, identify the trees to bring down, then you have to buck the logs, bring them on-site, split them, stack them, while also being careful to monitor the wood stove as it heats." And, as it turns out, if you want to enjoy cleared fields surrounding your house, "you have to mow . . . a lot."

■ ■ ■

Pete and Liz emphasize a perhaps surprising observation: when individuals in the FI community are provided large amounts of leisure time, they often voluntarily fill these hours with strenuous activity. This bias toward action over more traditional ideas of relaxation might strike some as needlessly exhausting, but to Pete and Liz it makes perfect sense.

Pete, for his part, offers three justifications for his strenuous life: it doesn't cost much money, it provides physical exercise, and it's good for his mental health ("For me, inactivity leads to a depressive boredom," he explains). Liz offers similar explanations for her decision to adopt the demands of rural living. She has a different name for these activities—"virtuous hobbies"—and emphasizes that activities that can seem like work actually offer multiple levels of benefits.

Consider, for example, the effort required to clear trails on their wooded property. As Liz told me: "We have property, we want to hike it, we have to clear trails to do this effectively, so we have to get out here with a chainsaw, cutting trees, clearing brush." This sounds like work, but it offers several different types of value. As Liz explained: "It is mentally freeing, because it is very different than working on a computer . . . it requires problem solving, but in a different way." In addition, it offers good exercise, and it requires you to learn new skills. "Learning to use a chainsaw is not easy," Liz told me. Finally, there's the satisfaction of actually getting to use the trail once cleared. As explained by Liz, a seemingly tedious task like clearing trails can suddenly seem significantly more rewarding than passively surfing Twitter.

The FI community, of course, is not the first to discover the inherent value in active leisure. Speaking to the Hamilton Club in Chicago in the spring of 1899, Theodore Roosevelt famously said: "I wish to preach, not the doctrine of ignoble ease, but the doctrine of the strenuous life." Roosevelt practiced what he preached. As president, Roosevelt regularly boxed (until a hard blow detached his left retina), practiced jujitsu, skinny-dipped in the Potomac, and read at the rate of one book per day. He was not one to sit back and relax.

A decade later, Arnold Bennett took up the cause of active leisure in his short but influential self-help guide, *How to Live on 24 Hours a Day*. In this book, Bennett notes that the average London middle-class white-collar worker putting in an eight-hour day is left with sixteen additional hours during which he is as free as any gentleman to pursue virtuous activity. Bennett argues that the waking half of these hours could be dedicated to enriching and demanding leisure, but were instead too often wasted by frivolous time-killing pastimes, like smoking, pottering, caressing the piano (but not actually playing), and perhaps deciding to become "acquainted with a genuinely good whiskey." After an evening of this mindless boredom busting (the Victorian equivalent of idling on your iPad), he notes, you fall exhausted into bed, with all the hours you were granted "gone like magic, unaccountably gone."

Bennett argues that these hours should instead be put to use for demanding and virtuous leisure activities. Bennett, being an early twentieth-century British snob, suggests activities that center on reading difficult literature and rigorous self-reflection. In a representative passage, Bennett dismisses novels because they "never demand any appreciable mental application." A good leisure pursuit, in Bennett's calculus, should require more "mental strain" to enjoy (he recommends difficult poetry). He also ignores the possibility that some of this leisure time might be reduced by childcare or housework, as he was writing only for men, who in Bennett's early twentieth-century middle-class British world, of course, never needed to bother with such things.

This is all to say, for our twenty-first-century purposes, we can ignore the specific activities Bennett suggests. What interests me instead is a more timeless piece of Bennett's argument, in which he fights the claim that his prescription of strained effort is too demanding to qualify as leisure:

What? You say that full energy given to those sixteen hours will lessen the value of the business eight? Not so. On the contrary, it will assuredly increase the value of the business eight. One of the chief things which my typical man has to learn is that the mental faculties are capable of a continuous hard activity; they do not tire like an arm or a leg. All they want is change—not rest, except in sleep.

This argument reverses our intuition. Expending more energy in your leisure, Bennett tells us, can end up energizing you more. He's reworking the old entrepreneurial adage "You have to spend money to make money" into the language of personal vitality.

This idea, which for lack of a better term we can call the *Bennett Principle*, provides a plausible foundation for the active leisure lives we've encountered so far in this section. Pete Adeney, Liz Thames, and Theodore Roosevelt all provide specific arguments for their embrace of strenuous leisure, but these arguments all build on the same general principle that the value you receive from a pursuit is often proportional to the energy invested. We might tell ourselves there's no greater reward after a hard day at the office than to have an evening entirely devoid of plans or commitments. But we then find ourselves, several hours of idle watching and screen tapping later, somehow more fatigued than when we began. As Bennett would tell you—and Pete, Liz, and Teddy would confirm—if you instead rouse the motivation to spend that same time actually doing something—even if it's hard—you'll likely end the night feeling better.

Pulling together these different strands, we identify our first lesson about cultivating high-quality leisure.

> **Leisure Lesson #1:** Prioritize demanding activity over passive consumption.

ON CRAFT AND SATISFACTION

Any conversation about high-quality leisure must eventually touch on the topic of craft. In this context, "craft" describes any activity where you apply skill to create something valuable. To make a fine table out of a pile of wood boards is an act of craft, as is knitting a sweater from a skein of yarn or renovating a bathroom without the help of contractors. Craft doesn't necessarily require that you create a new object, it can also apply to high-value *behaviors*. Coaxing a pleasing song out of a guitar or dominating a game of pickup basketball also qualifies. These definitions of craft can also apply to the digital world, where activities like computer programming or video gaming similarly require skill, but we should put an asterisk next to this final category for now—we'll return to it soon and unpack some of its complexities.

My core argument is that craft is a good source of high-quality leisure. Fortunately, when it comes to supporting this

argument, treatises on the value of craft are numerous—
starting with John Ruskin and the Arts and Crafts movement, and continuing through the modern maker community,
there have been thousands of books and articles written on the
topic. For our narrow purposes, a good starting place is Gary
Rogowski, a furniture maker based in Portland, Oregon. In
2017, Rogowski published a book titled *Handmade*, which is
part craftsman memoir and part philosophical investigation of
craft itself. What makes *Handmade* particularly relevant to our
discussion is that Rogowski specifically investigates the value
of craft in contrast to the lower-skilled digital behaviors that
dominate so much of our time—a purpose revealed by his
book's subtitle: *Creative Focus in the Age of Distraction*.

Rogowski provides several arguments for the value of craft
in a world increasingly mediated by screens, but I want to underscore one of these arguments in particular: "People have
the need to put their hands on tools and to make things. We
need this in order to feel whole." As Rogowski explains: "Long
ago we learned to think by using our hands, not the other way
around." As our species evolved, in other words, we did so as
beings that experience and manipulate the world around us.
We are orders of magnitude better at doing this than any
other animal, and this is true due to complex structures that
evolved in our brains to support this ability.

Today, however, it's easier than ever before to power down
these circuits. "Many people experience the world largely
through a screen now," Rogowski writes. "We live in a world that
is working to eliminate touch as one of our senses, to minimize

the use of our hands to do things except poke at a screen." The result is a mismatch between our equipment and our experience. When you use craft to leave the virtual world of the screen and instead begin to work in more complex ways with the physical world around you, you're living truer to your primal potential. Craft makes us human, and in doing so, it can provide deep satisfactions that are hard to replicate in other (dare I say) less *hands-on* activities.

The philosopher-mechanic Matthew Crawford is another useful source of wisdom on the value of craft-based leisure. After earning a PhD in political philosophy from the University of Chicago, Crawford took a quintessential knowledge-work job, running a think tank in Washington, DC. He soon grew disenchanted with the oddly disembodied and ambiguous nature of this work, so he did something extreme: he quit to start a motorcycle repair business. He now alternates between building custom motorcycles in his garage in Richmond, Virginia, and writing philosophical tracts on meaning and value in the modern world.

From his unique vantage as someone who has spent time working in both virtual and physical spaces, Crawford is particularly eloquent in describing the unique satisfactions of the latter:

> They seem to relieve him of the felt need to offer chattering *interpretations* of himself to vindicate his worth. He can simply point: the building stands, the car now runs, the lights are on. Boasting is what a boy

does, who has no real effect in the world. But craftsmanship must reckon with the infallible judgment of
reality, where one's failures or shortcomings cannot
be interpreted away.

In a culture where screens replace craft, Crawford argues,
people lose the outlet for self-worth established through unambiguous demonstrations of skill. One way to understand
the exploding popularity of social media platforms in recent
years is that they offer a substitute source of aggrandizement.
In the absence of a well-built wood bench or applause at a
musical performance to point toward, you can instead post
a photo of your latest visit to a hip restaurant, hoping for
likes, or desperately check for retweets of a clever quip. But
as Crawford implies, these digital cries for attention are often
a poor substitute for the recognition generated by handi-
craft, as they're not backed by the hard-won skill required to
tame the "infallible judgment" of physical reality, and come
across instead as "the boasts of a boy." Craft allows an escape
from this shallowness and provides instead a deeper source
of pride.

With these advantages established, we can now return to
our earlier asterisk on the claim that purely digital activities
can also be considered craft. There's clearly an argument to be
made that skilled digital behaviors generate satisfaction. I
made this point in my book *Deep Work*, where I noted that a
deep activity like writing a piece of computer code that solves

a problem (a high-skill effort) yields more meaning than a shallow activity like answering emails (a low-skill effort).

This being said, however, it's also clear that the specific benefits of craft cited here are grounded in their connection to the physical. While it's true that a digital creation can still generate the pride of accomplishment, both Rogowski and Crawford imply that activities mediated through a screen exhibit a fundamentally different character than those embodied in the real world. Computer interfaces, and the increasingly intelligent software running behind the scenes, are designed to eliminate both the rough edges and the possibilities inherent in directly confronting your physical surroundings. Typing computer code into an advanced integrated development environment is not quite the same as confronting a plank of maple wood with a handheld plane. The former misses both the physicality and sense of unlimited options latent in the latter. Similarly, composing a song in a digital sequencer misses the pleasures that come from the nuanced struggle between fingers and steel strings that defines playing a guitar well, while fast twitching your way to victory in *Call of Duty* misses many dimensions—social, spatial, athletic—present in a competitive game of flag football.

Because this chapter is about leisure—that is, efforts you voluntarily undertake in your free time—I'm going to propose that we stick to the stricter definition of craft promoted by the above arguments. If you want to fully extract the benefits of this craft in your free time, in other words, seek it in its analog forms, and while doing so, fully embrace Rogowski's closing advice: "Leave good evidence of yourself. Do good work." This then provides our second lesson about cultivating a high-quality leisure life.

> **Leisure Lesson #2:** Use skills to produce valuable things in the physical world.

SUPERCHARGED SOCIALITY

Another common property of high-quality leisure is its ability to support rich social interactions. Journalist David Sax witnessed the power of this property firsthand when an unusual café named Snakes & Lattes opened down the street from his Toronto apartment. This café didn't serve alcohol and offered no Wi-Fi, the food was forgettable and the chairs uncomfortable, and it cost five dollars just to enter. But as Sax reports in his 2016 book, *The Revenge of Analog,* on weekends the café's 120 seats would easily fill, with the line to enter spilling out onto the sidewalk. The wait for a table could be up to three hours.

The secret to Snakes & Lattes' success is that it's a *board game* café: you enter with a group of friends, are assigned a table, and then can select any game you want to play from the café's extensive library. If you need help, a game sommelier can make recommendations. The success of this café is somewhat puzzling, as analog games were supposed to disappear in a digital world. Why would you push plastic trinkets on a piece of cardboard when you could fight photorealistic ogres in a multiplayer video game like *World of Warcraft?* But they haven't. People are more eager than ever before to play Scrabble with neighbors, or trash-talk co-workers over poker, or line up in the Toronto cold for a table at Snakes & Lattes. The classic games that were popular in the pre-digital 1980s—Monopoly, Scrabble—remain popular sellers today, while the internet is fueling innovations in new game design (one of the most popular categories on Kickstarter is board games), leading to a renaissance in smarter, European-style strategy games—a movement best exemplified by the megahit Settlers of Catan, which has sold more than 22 million copies worldwide since it was first published in Germany in the mid-1990s.

David Sax argues that this popularity is due in large part to the social experience of playing these games. "Tabletop gaming creates a unique social space apart from the digital world," he writes. "It is the antithesis of the glossy, streaming waterfalls of information and marketing that masquerade as relationships on social networks." When you sit down at a table to play a game in person with other people, you're exposing yourself to what the game theorist Scott Nicholson calls "a rich multimedia, 3D interaction." You scrutinize your opponent's body language in search of clues about their strategy and try to project yourself into their mind to understand what they might be plotting in their next moves, searching for what Sax calls "the signal flares of our most complex emotions."

The sting of defeat is all the more real when you sit across from your smiling victor while packing up the pieces, but because the defeat is within the structured confines of a game, it fades, allowing you to practice the complex inter-social dance required to defuse the tension. We're wired for these master-level social chess matches, and games allow us to push these abilities to their limits—a thrilling experience.

Playing games also provides permission for what we can call *supercharged socializing*—interactions with higher intensity levels than are common in polite society. Sax describes the excited chatter and loud belly laughs he encountered at Snakes & Lattes during a busy night. This observation doesn't surprise me. Every couple of months, a group of dads I know get together to (poorly) play poker. These sessions provide us an excuse to joke and chat and vent for three hours. When a player in our game runs out of chips early, he always sticks around for the rest of the game. It's not really about the cards, just as playing Catan at Snakes & Lattes is not really about building roads.

These benefits of old-fashioned, in-person playing help explain why even the fanciest video games and shiniest mobile entertainments haven't ruined the board game industry. As Sax writes: "On a social level, video games are decidedly low bandwidth compared to the experience of playing a game on a square of flat cardboard with another human being."

Board games, of course, are not the only type of leisure that promote intense social experiences. Another interesting intersection of leisure and interaction is emerging in the world of health and exercise. Arguably one of the biggest trends in this sector is the "social fitness" phenomenon, in which, as one sports industry analyst describes it, "fitness has shifted from a private activity at the gym to a social interaction in the studio or on the street."

If you live in a city, you've probably seen groups who gather in the park to be put through boot-camp-style calisthenics by a barking instructor. The group I used to see gather on a grass patch near my local Whole Foods consists of new mothers who arrange themselves in a ring around their strollers. I don't know if this group offers better fitness results than the Planet Fitness gym that operates a few blocks down the road from this location, but the social experience is almost certainly much richer. To meet with the same group of women, who are all facing the same challenges of new motherhood, enables a level of interaction and support that's entirely missing when you walk into a fluorescent-lit gym with your earbuds blaring.

Another popular group fitness organization is F3, which stands for Fitness, Fellowship and Faith. F3 is only for men and is entirely volunteer led, with no money ever charged. The concept is that you join or start a local group that meets several times a week for an outdoor workout—rain or shine. Given that the workout leader is a position that rotates among the group members, men aren't drawn to F3 for expert fitness guidance. They're attracted to the social experience. This reality is evidenced by the almost comic level of male camaraderie that members embrace (with a knowing nod). As the F3 website explains:

> For FNGs [new members], the swirl of inside-baseball lingo and jargon used at your average F3 workout can be a bit confusing. Like, for instance, What's an FNG and why do people keep calling me that?

The site then provides a "lexicon" of F3 jargon that contains over a hundred different alphabetized entries, many of which reference other entries, creating a complex recursive morass. Case in point, the following definition from the lexicon:

> BOBBY CREMINS (as in, to pull one): When a man Posts to one Workout, but leaves after Startex to go to a different AO. Also, a non-Workout LIFO initiated by the M or CBD.

To an FNG like me, this definition makes no sense. But then again, that's the point. By the time you *do* understand what it means to pull a Bobby Cremins, you'll have earned a satisfying sense of having been accepted by a tribe. This pursuit of inclusion is perhaps best exemplified by the circle-of-trust ritual that ends each workout. During the ritual, each participant gives their own name and their F3 nickname before offering some words of wisdom or gratitude. If you're new to the group, you're given a nickname on the spot—an initiation.

To some, these artificial rules and jargon might seem a little over the top, but its effectiveness is undeniable. The first free F3 workout was led by co-founders David Redding (nickname "Dredd") and Tim Whitmire (nickname "OBT") on the campus of a Charlotte-area middle school in January 2011. Seven years later, there are over 1,200 groups operating around the country.

The biggest success story from the social fitness phenomenon, however, is unquestionably CrossFit. The first CrossFit gym (called a "box" in CrossFit jargon) opened in 1996. There are now more than 13,000 boxes in over 120 countries. In the US, there's one CrossFit box for every two Starbucks—an incredible reach for a fitness brand.

When first encountered, CrossFit's popularity confused industry insiders who for years had focused relentlessly on price and services at their gyms. The typical CrossFit box is a somewhat grimy, largely empty warehouse. The fitness equipment— often pushed to the peripheries—would fit in well in a turn-of-the-century boxing gym: kettlebells, medicine balls, ropes, wooden boxes, pull-up bars, and metal squat racks. You won't find treadmills, fancy cable machines, nice locker rooms, bright lights, or, God forbid, television screens. It's also really expensive. The Planet Fitness near my house costs $10 a month—a price that includes free Wi-Fi. The CrossFit box near my house costs $210 a month, and if you ask them about Wi-Fi, they'll chase you out the door with a kettlebell.

The secret to CrossFit's success is probably best captured by one of the most notable differences between a CrossFit box and a standard gym: no one is wearing earphones. The CrossFit fitness model is built around the workout of the day (or WOD)—which is typically a high-intensity combination of functional movement exercises that you try to execute as quickly as possible. Here's a sample WOD from around the time I was first writing this chapter:

3 rounds for time of:

- 60 squats
- 30 knees-to-elbows
- 30 ring push-ups

You're not allowed to do the WOD on your own. There are instead a small number of preselected times each day during which you can show up at your local box and execute the WOD along with a group of other members and a supervising trainer. The social aspect of the workout is crucial: you cheer on the group while they in turn cheer you on. This support helps push people past their natural limits, which is important; a core belief of CrossFit is that extreme intensity in a short period of time is superior to a large volume of exercise over a long period. The social aspect of the WOD also helps create a strong sense of community. Here's how a former personal trainer turned CrossFit devotee describes the experience: "The camaraderie of other members cheering me on to finish strong as I fought for a few more reps during a WOD at [my CrossFit box] was an exhilarating feeling which I never have experienced at any other fitness facility." Greg Glassman, CrossFit's outspoken founder, captures the sense of rough-edged but intense camaraderie created by his fitness movement by famously describing CrossFit as a "religion run by a biker gang."

■   ■   ■

The local new-mom boot camp, F3, and CrossFit are successful for the same reason as the Snakes & Lattes board game café: they are leisure activities that enable the types of energized and complex sociality that are otherwise rare in normal life. Board games and social fitness are not the only leisure activities that can generate these social benefits. Other examples include recreational sports leagues, most volunteer activities, or working with a team on a group project, like fixing up an old boat or building a neighborhood skating rink.

The most successful social leisure activities share two traits. First, they require you to spend time with other people in person. As emphasized, there's a sensory and social richness to real-world encounters that's largely lost in virtual connections, so spending time with your *World of Warcraft* clan doesn't qualify. The second trait is that the activity provides some sort of structure for the social interaction, including rules you have to follow, insider terminology or rituals, and often a shared goal. As argued, these constraints paradoxically enable more freedom of expression. Your CrossFit buddies will holler and whoop, and give you emphatic high fives and sweaty hugs with a joyous enthusiasm that would seem insane in most other contexts.

We can now conclude this exploration by stating our third lesson about cultivating a high-quality leisure life.

> **Leisure Lesson #3:** Seek activities that require
> real-world, structured social interactions.

THE LEISURE RENAISSANCE

The Mouse Book Club provides a good example of the complex relationship between high-quality leisure and digital technology. If you join this club, you will receive, four times a year, a themed collection of classic books and short stories. The collection released during the 2017 holiday season, for example, followed a "giving" theme and included "The Gift of the Magi," by O. Henry, "The Happy Prince," by Oscar Wilde, and a collection of three Russian Christmas stories, penned by Tolstoy, Dostoevsky, and Chekhov.

What differentiates this club from similar organizations is the books themselves, which are custom printed in a compact booklet that's roughly the height and width of a smartphone. This size is intentional. The philosophy behind a Mouse Book is that it can fit into your pocket next to your phone. Whenever you feel the urge to pull out your phone for a quick hit of distraction, you can instead pull out the Mouse Book and read a few pages of something deeper. The company describes their goal as "mobilizing literature," and likes to point out that their portable entertainment devices "never run out of battery life, their 'screens' never crack, and they don't ring, buzz, or vibrate."

Like the other examples of high-quality leisure highlighted in this chapter, a Mouse Book is defiantly analog. It's a physical object that demands (cognitive) struggle before it begins to return value—but when it does, the value is more substantial and lasting than the sugar high of a lightweight digital distraction. These examples can seem to place high-quality leisure into an antagonistic relationship with newer technologies, but as I hinted above, the reality is more complicated. A closer look at the Mouse Book Club makes clear that its existence depends on multiple technological innovations.

Printing books requires capital. The project's co-founders, David Dewane and Brian Chappell, raised this money with an online Kickstarter campaign that attracted over $50,000 in funding from more than 1,000 backers. These backers found their way to this campaign in part because of bloggers like me who directed their online followings toward the project. Another key aspect of the Mouse Book Club model is helping readers understand and discuss the books they're sent, enabling them to maximize the value they receive from their reading experience. To do so, the company launched a blog that allows their editors to discuss the themes from the latest collection, and started an interview-based podcast to dive into select ideas. (The most recent episode is an interview about Montaigne with Philippe Desan, a respected literature professor from the University of Chicago.) As I write this chapter, the company is also in the process of building an online system to help nearby subscribers find each other and organize real-world book club meetings.

The Mouse Book Club delivers a high-quality analog experience, but it couldn't exist without many technological innovations of the past decade. I'm pointing this out to push back on the idea that high-quality leisure requires a nostalgic turning back of time to a pre-internet era. On the contrary, the internet is fueling a *leisure renaissance* of sorts by providing the average person more leisure options than ever before in human history. It does so in two primary ways: by helping people find communities related to their interests and providing easy access to the sometimes obscure information needed to support specific quality pursuits. If you move to a new city and want to find other people who share your interest in debating literature, the Mouse Book Club can help connect you to some nearby bibliophiles. If, inspired by the Frugalwoods blog, you want to start gathering your own firewood, there are any number of YouTube videos that can teach you the basics. I can't think of a better time than the present to cultivate a high-quality leisure life.

We've now arrived at an apparent circularity. This chapter argues that to escape the drain of low-value digital habits, it's important to first put in place high-quality leisure activities. These quality activities fill the void your screens were previously tasked to help you ignore. But I just argued that you should use digital tools to help cultivate this leisure. It seems, then, that I'm asking you to embrace new technology to help you avoid new technology.

Fortunately, this circularity is easily broken. The state I'm helping you escape is one in which passive interaction with your screens is your primary leisure. I want you to replace this with a state where your leisure time is now filled with better pursuits, many of which will exist primarily in the physical world. In this new state, digital technology is still present, but now subordinated to a support role: helping you to set up or maintain your leisure activities, but not acting as the primary source of leisure itself. Spending an hour browsing funny You-Tube clips might sap your vitality, while—and I'm speaking from recent experience here—using YouTube to teach yourself how to replace a motor in a bathroom ventilation fan can provide the foundation for a satisfying afternoon of tinkering.

A foundational theme in digital minimalism is that new technology, when used with care and intention, creates a better life than either Luddism or mindless adoption. We shouldn't be surprised, therefore, that this general idea applies here to our specific discussion of cultivating leisure.

■ ■ ■

Aristotle argued that high-quality leisure is essential to a life well lived. With this in mind, in this chapter I provided three lessons about how to cultivate these high-quality pursuits. I then concluded with the caveat that although these activities are primarily analog in nature, their successful execution often depends on the strategic use of new technologies.

As with the other chapters in part 2 of this book, I'll conclude our discussion of leisure with a collection of concrete practices that can help you act on these insights. These practices do not constitute a step-by-step plan for upgrading your leisure life, but instead provide a sampling of the *type* of action that can help you operationalize Aristotle's blueprint for happiness.

PRACTICE: FIX OR BUILD SOMETHING EVERY WEEK

Earlier in this chapter, I introduced Pete Adeney (a.k.a. Mr. Money Mustache), the former engineer who achieved financial independence at a young age. If you sift through the archive of Pete's blog, you might come across a remarkable entry from April 2012, which describes Pete's experiments with metal welding.

As Pete explains, his welding odyssey began in 2005. At the time, he was building a custom home. (As loyal Mr. Money Mustache fans know, Pete spent a few years running a somewhat ill-fated home construction company after quitting his job as an engineer.) The house was modern so Pete integrated some custom metalwork into his design plan, including a beautiful custom steel railing on the stairs.

The design seemed like a great idea until Pete received a quote from his metal contractor for the work: it was for $15,800, and Pete had budgeted only $4,000. "Damn! . . . If this guy is billing out his metalworking time at $75.00 an hour, that's a sign that I need to finally learn the craft myself," Pete recalls thinking at the time. "How hard can it be?" In Pete's hands, the answer turned out to be: *not that hard.*

As he details in his post, Pete bought a grinder, a metal chop saw, a visor, heavy-duty gloves, and a 120-volt wire-feed flux core welder—which, as Pete explains, is by far the easiest welding device to learn. He then picked some simple projects, loaded up some YouTube videos, and got to work. Before long, Pete became a competent welder—not a master craftsman, but skilled enough to save himself tens of thousands of dollars in labor and parts. (As Pete explains it, he can't craft a "curvaceous supercar," but he could certainly weld up a "nice Mad-Max-style dune buggy.") In addition to completing the railing for his custom home project (for much less than the $15,800 he was quoted), Pete went on to build a similar railing for a rooftop patio on a nearby home. He then started creating steel garden gates and unusual plant holders. He built a custom lumber rack for his pickup truck and fabricated a series of structural parts for straightening up old foundations and floors in the historic homes in his neighborhood. As Pete was writing his post on welding, a metal attachment bracket for his garage door opener broke. He easily fixed it.

Pete is an example of someone who is *handy*, in the sense that he's comfortable picking up a new physical skill when needed. There was a time in this country when most people were handy. If you lived in a rural area, for example, you had to be comfortable fixing and building things—there was no Amazon Prime to deliver a replacement or Yelp-approved contractor to stop by with his tools. Matthew Crawford points out that the Sears catalog used to include blown-up parts diagrams for all of their appliances and mechanical goods. "It was simply taken for granted that such information would be demanded by the consumer," he writes.

Handiness is rarer today for the simple reason that, for most people, it's no longer essential for either their professional or home lives to function smoothly. This transition has pros and cons. The main pro, of course, is that it frees up massive numbers of hours to be put toward more productive use. There's a thrill to fixing something that's broken, but if you're *constantly* fixing things, it can get old. Economists will also argue that specialization is more efficient. If you're a lawyer, you're better off, from a financial perspective, dedicating your time to becoming a better lawyer, and then trading some of the extra money you earn to people who specialize in fixing when something breaks.

But maximizing personal and financial efficiency isn't the only relevant goal. As I argued earlier in this chapter, learning and applying new skills is an important source of high-quality leisure. If you can achieve some degree of handiness, therefore, you can more easily tap into this type of satisfying activity. This practice won't ask you to become Pete Adeney—who, as we previously explored, has near endless time for such pursuits—but it will push you to make straightforward repair, learning, or building projects a regular part of your routine.

■ ■ ■

The simplest way to become more handy is to learn a new skill, apply it to repair, learn, or build something, and then repeat. Start with easy projects in which you can follow

step-by-step instructions more or less directly. Once comfortable, advance toward more-complicated endeavors that require
you to fill in some blanks or adapt what's suggested. To be
more concrete, here's a sample list of the types of straightforward projects I had in mind for someone new to using their
hands for useful purposes. Every example below is something
that either I or someone I know was able to learn and execute
in a single weekend.

- Changing your own car oil
- Installing a new ceiling-mounted light fixture
- Learning the basics of a new technique on an instru-
  ment you already play (e.g., a guitar player learning Tra-
  vis picking)
- Figuring out how to precisely calibrate the tone arm on
  your turntable
- Building a custom headboard from high-quality lumber
- Starting a garden plot

Notice that none of these projects are digital. Though there
is some pride to be gained in learning a new computer program, or figuring out a complicated new gadget, most of us
already spend enough time moving symbols around on screens.
The leisure we're tackling here is meant to tap into our strong
instinct for manipulating objects in the physical world.

If you're wondering *where* to learn skills needed for simple
projects like those listed above, the answer is easy. Almost
every modern-day handyperson I've spoken to recommends

the exact same source for quick how-to lessons: YouTube. For any standard project, there are numerous YouTube videos to walk you through the process. Some are more informative than others, but as you become more confident, you won't need precise instructions—steps that point you in the generally right direction will be enough.

My suggestion is that you try to learn and apply one new skill every week, over a period of six weeks. Start with easy projects like those suggested above, but as soon as you feel the challenge wane, ramp up the complication of the skills and steps involved.

When this six-week experiment ends, you won't quite be ready to rebuild the engine on your Honda, but you'll have achieved entry-level handy status. That is, just enough competence to realize you're capable of learning new things, and to realize that you enjoy doing so. If you're like most, this six-week crash course will spark a persistent and rewarding inclination toward getting your hands dirty.

PRACTICE: SCHEDULE YOUR LOW-QUALITY LEISURE

A few years ago, the Silicon Valley business pioneer Jim Clark was interviewed at an event held at Stanford University. At some point in the interview, the topic turned to social media. Clark's reaction was unexpected given his high-tech background: "I just don't appreciate social networking." As he then clarifies, this distaste is captured by a particular experience he had sitting on a panel with a social media executive:

> [The executive was] just raving about these people spending twelve hours a day on Facebook . . . so I asked a question to the guy who was raving: "The guy who's spending twelve hours a day on Facebook, do you think he'll be able to do what you've done?"

In this question, Clark puts his finger on the central flaw afflicting the utopian vision promoted by Web 2.0's biggest boosters. Tools like Facebook and Twitter are marketed in terms of the positive things they can enable, such as connection and expression. But as revealed in the enthusiasm of Clark's fellow panel member, to the large attention economy conglomerates, these benefits are like the prize in the Cracker Jack box—something appealing to get you to tap the app, at which point they can proceed with their primary objective of extracting as many minutes of your time and attention as possible for their profit machine. (See part 1 for a more detailed discussion of the psychological vulnerabilities these services exploit to succeed in this goal.)

As Clark incredulously pointed out, no matter what immediate benefits these services might provide the users, the net impact on their productivity and life satisfaction must be profoundly negative if *all* these users do is engage the service. You can't, in other words, build a billion-dollar empire like Facebook if you're wasting hours every day using a service like Facebook.

This tension between the benefits provided by the attention economy and this sector's primary mission of devouring your time proves particularly problematic for our current goal of cultivating high-quality leisure. It's too easy to be good intentioned about adding some quality activity into your evening, and then, several hours of rabbit hole clicking and binge-watching later, realize that the opportunity has once again dissipated.

A straightforward solution to this problem would be to stop using most of these engineered distractions. As you dive deeper into the minimalism philosophy taught in this book, this might be exactly what you end up doing. But this drastic step is getting ahead of ourselves. The premise of this chapter is that by cultivating a high-quality leisure life *first*, it will become easier to minimize low-quality digital diversions *later*. With this in mind, I want to offer a simpler solution, one that doesn't yet require you to seriously cull the services and sites you frequent, but that will nonetheless make it easier for you to put aside time for quality leisure. It also has the advantage, as I'll soon elaborate, of being an idea that terrifies social media companies.

■　■　■

Here's my suggestion: *schedule in advance the time you spend on low-quality leisure.* That is, work out the specific time periods during which you'll indulge in web surfing, social media checking, and entertainment streaming. When you get to these periods, anything goes. If you want to binge-watch Netflix while live-streaming yourself browsing Twitter: go for it. But outside these periods, stay offline.

There are two reasons why this strategy works well. First, by confining your use of attention-capturing services to well-defined periods, your remaining leisure time is left protected for more substantial activities. Without access to your standard screens, the best remaining option to fill this time will be quality activities.

The second reason this strategy works well is that it doesn't ask you to completely abandon low-quality diversions. Abstention activates subtle psychologies. If you decide, for example, to avoid all online activities during your leisure time, this might generate too many minor issues and exceptions. The part of your mind that is skeptical of your newfound enthusiasm for disconnection will use these objections to undermine your determination. Once undermined, your commitment to restriction will crumble and you'll be thrown back into a state of unrestricted and compulsive use.

On the other hand, if you're simply corralling these behaviors to specific periods, it becomes much harder for the skeptical part of your mind to mount a strong case. You're not quitting anything or losing access to any information, you're simply being more mindful of when you engage with this part of your leisure life. It's difficult to paint such a reasonable restriction as untenable, which makes it more likely to last.

When first implementing this strategy, don't worry about *how much* time you put aside for low-quality leisure. It's fine, for example, if you start with major portions of your evenings and weekends dedicated to such pursuits. The aggressiveness of your restrictions will naturally increase as they allow you to integrate more and more higher-quality pursuits into your life.

The element of this practice that terrifies social media companies is that you'll learn through experience that even after you significantly reduce the time you spend on these services, you won't feel like you're missing many benefits. I conjecture that the vast majority of regular social media users can receive the vast majority of the value these services provide their life in as little as twenty to forty minutes of use *per week*. This is why even serious constraints to your schedule won't lead you to feel like you're missing out on something important. This observation terrifies social media companies because their business model depends on your engaging their products for as many minutes as possible. This is why, when defending their products, they prefer to focus on the question of *why* you use them, not *how* you use them. Once people start thinking seriously about the latter question, they tend to recognize that they're spending way too much time online. (I'll dive deeper into this issue in the next chapter.)

These reasons help explain the surprising effectiveness of this simple strategy. Once you start constraining your low-quality distractions (with no feeling of lost value), and filling the newly freed time with high-quality alternatives (which generate significantly higher levels of satisfaction), you'll soon begin to wonder how you ever tolerated spending so many of your leisure hours staring passively at glowing screens.

PRACTICE: JOIN SOMETHING

Benjamin Franklin, who was naturally gregarious, instinctually understood the argument I made earlier about the importance of structured social interactions. Acting on this instinct, however, required hard work for this future founding father. When Franklin returned from London to Philadelphia in 1726, he faced a barren social life. Having grown up in Boston, Franklin had no family roots in his adopted home, and his skepticism of religious dogma eliminated the option of joining a ready-made community through the church. Undeterred, he decided he would simply start the social organizations he desired from scratch.

In 1727, Franklin created a social club called the Junto, which he describes as follows in his autobiography:

> I had form'd most of my ingenious acquaintance into a club of mutual improvement, which we called the Junto; we met on Friday evenings. The rules that I drew up required that every member, in his turn, should produce one or more queries on any point of

Morals, Politics, or Natural Philosophy, to be discuss'd
by the company; and once in three months produce
and read an essay of his own writing, on any subject he
pleased.

Inspired by these meetings, Franklin created a scheme in
which the Junto members would contribute funds toward buying books that all members could use. This model soon grew
beyond Franklin's Friday evening gatherings, leading him in
1731 to write the charter for the Library Company of Philadelphia, one of the first subscription libraries in America.

In 1736, Franklin organized the Union Fire Company, one
of the first volunteer firefighting companies in America and a
much-needed service given the flammability of colonial-era
cities. By 1743, as his interest in science grew, Franklin organized the American Philosophical Society (which still exists
today) as a more efficient way to connect the smartest scientific minds in the country.

These efforts in creating new social organizations also
succeeded in gaining him the contacts needed to access long-
existing clubs. To name a notable example, Franklin was invited in 1731 to join the local Masonic lodge. By 1734, he'd
risen to the rank of grand master—a fast rise that underscores
his dedication to the group.

Perhaps most amazingly, all of this social activity took place
before his retirement from the printing business in 1747,
which, in Franklin's recounting, was the turning point after
which he could *finally* get serious about his leisure time.

■ ■ ■

Franklin is one of the great socializers in American history. His commitment to structured activities and interactions with other people provided this restless founder great satisfaction and, more pragmatically speaking, built the foundation for his successes in business and then, later, politics. Few can mimic the energy Franklin invested into his social leisure, but we can all extract an important lesson from his approach to cultivating a fulfilling leisure life: *join things.*

Franklin was relentlessly driven to be part of groups, associations, lodges, and volunteer companies—any organization that brought interesting people together for useful ends captured his attention as a worthwhile endeavor. As we have seen, when he couldn't find such gatherings, he created them from scratch. This strategy worked. He arrived in Philadelphia an unknown. Two decades later he had risen to become one of its most connected and respected citizens, as well as one of its most engaged. Listlessness and boredom were not common companions in Franklin's frenetic life.

We would do well to keep in mind Franklin's lesson about joining. It's easy to get caught up in the annoyances or difficulties inherent in any gathering of individuals struggling to work toward a common goal. These obstacles provide a convenient excuse to avoid leaving the comfort of family and close friends, but Franklin teaches us that it's worth pushing past these concerns. Join first, he would advise, and work out the other issues later. It doesn't matter if it's a local sporting league, a committee at your temple, a local volunteer group, the PTA, a social fitness group, or a fantasy gamers club: few things can replicate the benefits of connecting with your fellow citizens, so get up, get out, and start reaping these benefits in your own community.

PRACTICE: FOLLOW LEISURE PLANS

In the professional world, many high achievers are meticulous strategists. They lay out a vision for what they're trying to accomplish on multiple different time scales, connecting high-level ambition to decisions about daily actions. I've both practiced and written about these types of professional strategies for many years.* Here I want to suggest that you apply this same approach to your leisure life. I want you, in other words, to strategize your free time.

If your leisure is dominated by low-quality activities, then the idea that you need a strategy might sound absurd—how much forethought is needed to support web surfing or binging on Netflix? But for those who embrace high-quality leisure, the benefits of a strategic approach are more obvious, as this class of activity often requires more-complicated scheduling

---

* For a representative sample of my thinking on this topic, browse my blog archive at calnewport.com/blog for numerous articles on weekly and daily planning. I also touch on these issues in detail in my previous book *Deep Work*.

and organization. Without a well-considered approach to your high-quality leisure, it's easy for your commitment to these pursuits to degrade due to the friction of everyday life.

With this in mind, I suggest you strategize this part of your life with a two-level approach consisting of both a *seasonal* and *weekly* leisure plan. I explain each below.

The Seasonal Leisure Plan

A seasonal leisure plan is something that you put together three times a year: at the beginning of the fall (early September), at the beginning of the winter (January), and at the beginning of summer (early May). I'm preferential toward seasonal timing as I'm an academic, and this matches the university calendar. Those with a business background might prefer quarterly planning, which works fine too. You can use whatever semiannual schedule seems most natural to you, but for simplicity I'll stick with the seasonal suggestion throughout this discussion.

A good seasonal plan contains two different types of items: *objectives* and *habits* that you intend to honor in the upcoming season. The objectives describe specific goals you hope to accomplish, with accompanying strategies for how you will accomplish them. The habits describe behavior rules you hope to stick with throughout the season. In a seasonal *leisure* plan, these objectives and habits will both be connected to cultivating a high-quality leisure life.

Here's an example of a well-crafted objective that you might find in a seasonal leisure plan:

Objective: Learn on the guitar every song from the A-side of *Meet the Beatles!*

Strategies:

- Restring and retune my guitar, find the chord charts for the songs, print them, and put them in nice plastic protector sheets.
- Return to my old habit of regularly practicing my guitar.
- As incentive, schedule Beatles party in November. Perform songs (get Linda to agree to sing).

Notice the use of specificity in the objective description. If our hypothetical leisure planner had instead written, "play guitar more regularly," she would have been less likely to succeed, as the goal is vague and too easy to ignore. She instead identified a concrete accomplishment that has clear criteria for completion and that can reasonably fit within a season. By pursuing this accomplishment, of course, she'll be forced to act on her vaguer commitment to play her guitar more regularly.

Also notice that the strategies for achieving the objective include an incentive: scheduling a party that will require her to have learned the songs. This isn't mandatory, but it's always helpful to give yourself a deadline when possible. Finally, notice that she doesn't get too specific about the scheduling details of the ongoing strategies. She notes she needs to regularly practice, but doesn't specify when she'll do this practice each week, or how long the sessions will last. The details of this scheduling are best left to the weekly planning process described below.

Moving on, here are several examples of the other type of item found on seasonal leisure plans, the habits:

Habit: During the week, restrict low-quality leisure to only sixty minutes a night.

Habit: Read something in bed every night.

Habit: Attend one cultural event per week.

Each of the habits describes an ongoing behavior rule. They're not dedicated to a particular objective, but instead are designed to maintain a background commitment to regular high-quality leisure in the planner's life.

The boundary between habits and objectives is porous. In our above examples, our hypothetical planner might have added "practice guitar twice a week" to her habit list instead of including it in her Beatles-themed objective. Similarly, she might have transformed her "read every night" habit into an objective about reading a specific group of books during the season, an objective that would end up requiring daily reading to accomplish.

This porousness is unavoidable in this exercise and should not be a major source of concern. A good seasonal plan will have a small number of interesting and motivating objectives, coupled with a small number of tractable habits designed to ensure a regular patina of quality. How you shift specific leisure ideas between these two categories is less important than keeping them reasonable and balanced for the season ahead.

The Weekly Leisure Plan

At the beginning of each week, put aside time to review your current seasonal leisure plan. After processing this information, come up with a plan for how your leisure activities will fit into your schedule for the upcoming week. For each of the objectives in the seasonal plan, figure out what actions you can do during the week to make progress on these objectives, and then, crucially, schedule exactly *when* you'll do these things.

Let's return to our above example about the Beatles-themed guitar objective. The weekly leisure plan is when you'll figure out how this practice will fit into your schedule. Let's say our hypothetical planner schedules the gym from 7:30 to 8:30 a.m. before work on Monday, Wednesday, and Friday. She might then decide in the upcoming week that she'll use this 7:30–8:30 slot for guitar practice on Tuesday and Thursday. Maybe on another week, however, a series of early morning meetings makes this timing unavailable. She might then identify some empty evenings for her weekly practice.

If you're already in the habit of creating detailed plans for your week (which I highly recommend), you can just integrate your weekly leisure plan into whatever system you already use for planning. The more you see these leisure plans as just part of your normal scheduling—and not some separate and potentially optional endeavor—the more likely you are to succeed in following them.

Finally, when you are done with this schedule, take time to review and remind yourself of the habits included in your seasonal plan. These reminders will prevent you from forgetting these commitments in the week ahead. It can also be useful to briefly reflect on your experience with the habits in the week that just ended. Some people like to keep simple scorecards throughout the week of how often they stuck with the rules specified by these habits, and review the scorecard as part of this reflection. The goal here is twofold. First, knowing that you will soon review your performance makes you more likely in the moment to stick with your habits. Second, this reflection allows you to identify issues that might need resolving. If you're consistently failing to execute a given habit, regardless of your efforts to cajole yourself into action, there might be an issue with the habit itself that makes it difficult to satisfy.

■　■　■

You might be concerned that injecting more systematic thinking into your leisure life will rob it of the spontaneity and relaxation you crave for the time left over after your professional and family obligations. I hope to convince you that this concern is overblown. The weekly leisure planning process itself requires only a handful of minutes, and scheduling in advance some high-quality leisure activities hardly removes all spontaneity from your free time.

In addition, I've noticed that once someone becomes more intentional about their leisure, they tend to find more of it in their life. The weekly planning ritual can lead you to begin fighting for more leisure opportunities. Seeing, for example, that Thursday is a light schedule, you might decide to end work at 3:30 that day to go on a hike before dinner. These types of invented opportunities are rarer when you're not planning ahead. Becoming more systematic about your leisure, in other words, can significantly increase the relaxation you enjoy throughout your week.

Finally, in justifying this planning approach, I want to underscore the foundational argument delivered throughout this chapter: doing nothing is overrated. In the middle of a busy workday, or after a particularly trying morning of childcare, it's tempting to crave the release of having *nothing* to do—whole blocks of time with no schedule, no expectations, and no activity beyond whatever seems to catch your attention in the moment. These decompression sessions have their place, but their rewards are muted, as they tend to devolve toward low-quality activities like mindless phone swiping and half-hearted binge-watching. For the many different reasons argued in the preceding pages, investing energy into something hard but worthwhile almost always returns much richer rewards.

Join the Attention Resistance

DAVID AND GOLIATH 2.0

In June of 2017, Facebook launched a blog series titled "Hard Questions." The announcement for this series, written by their vice president for public policy and communications, admitted that as "digital technologies transform how we live, we all face challenging questions." The series, he explained, would be a chance for Facebook to explain how they are grappling with these questions.

In the period between that initial announcement and the winter of 2018, Facebook published fifteen articles, tackling a variety of topics. In June, they explored the issues surrounding the identification of hate speech in a global community. In September and October, they discussed the Russian Facebook ads that played a role in the 2016 presidential election. In December, they pushed back on general fears surrounding facial recognition technology, which Facebook uses for purposes like auto-tagging photos. "Society often welcomes the benefit of a new innovation while struggling to harness its potential," they wrote, before helpfully noting that in 1888 some people were worried about Kodak cameras.

At the time, I tepidly applauded Facebook for being more open about their thinking on these questions, but for the most part wasn't that interested in this corporate communication exercise. That is, until they published an article tackling a more significant prompt: "Is Spending Time on Social Media Bad for Us?" Written by two Facebook researchers named David Ginsberg and Moira Burke, this article, which we briefly touched on in an earlier chapter when we discussed what science teaches us about social media's harm and benefits, opens with the observation that "a lot of smart people are looking at different aspects of this important issue." Taking advantage of this reality, the authors then survey the academic literature for more clarity on what are the "good" and "bad" ways to engage with social media, concluding: "According to the research, it really comes down to how you use the technology."

As I'll argue, this post represented a momentous shift in how Facebook talks about itself—a shift that might turn out to be a major folly for the social media giant, and perhaps even mark the beginning of the end of its current moment of cultural ubiquity. More importantly, as I'll show, it inadvertently reveals an effective strategy for maintaining your autonomy in a period when numerous digital forces want to diminish it.

■ ■ ■

To understand my claim about Facebook's folly, we must first step back to understand the attention economy in which it operates. It's important to know that the "attention economy" describes the business sector that makes money gathering consumers' attention and then repackaging and selling it to advertisers. This idea is not new. Columbia Law professor and technology scholar Tim Wu (who wrote a book on this topic titled *The Attention Merchants*) traces the beginning of this economic model to 1830, when the newspaper publisher Benjamin Day launched the *New York Sun*, the first penny press newspaper.

Up to that point, publishers considered their readers to be their customers, and saw their goal as providing a product good enough to convince people to pay to read it. Day's innovation was to realize that his readers could become his product and the advertisers his customers. His goal became to sell as many minutes of his readers' attention as possible to the advertisers. To do so, he lowered the price of the *Sun* to a penny and pushed more mass interest stories. "He was the first person to really appreciate the idea—you gather a crowd, and you're not interested in that crowd for its money," Wu explained in a speech, "but because you can resell them to someone else who wants their attention."

This business model caught on, sparking the tabloid wars of the nineteenth century. It was then adopted by the radio and television industries in the twentieth century, where it was pushed to new extremes as these emerging mass media technologies were wielded to gather crowds of unprecedented size.

Not surprisingly, once the consumer internet went mainstream in the late 1990s, there was a scramble to figure out how to adapt this model to the online world. Initial attempts were not that successful (think: pop-up ads). In the mid-2000s, when Google went public, it was valued at a modest $23 billion. The most expensive internet company at the time was eBay, which made its money from commissions and was worth only about twice that. Facebook existed, but it was still called thefacebook.com and was open only to college students.

A decade later, this has all changed. During the week when I'm writing these words, Google is the second most valuable company in the United States, with a market cap of over $800 billion. Facebook, which had fewer than a million users ten years ago, now has over two billion and is the fifth most valuable company in the US, with a market cap of over $500 billion. ExxonMobil, by contrast, is currently worth around $370 billion. Extracting *eyeball minutes*, the key resource for companies like Google and Facebook, has become significantly more lucrative than extracting oil.

To understand how this massive change occurred, you need look no further than the number one largest company in the country: Apple. The iPhone, and the imitators that soon followed, enabled the attention economy to shift from its historical position as a profitable but somewhat niche sector to one of the most powerful forces in our economy. At the core of this shift was the smartphone's ability to deliver advertisements to users at all points during their day, as well as to help services gather data from these users to target those advertisements with unprecedented precision. It turns out that there remained vast reservoirs of human attention that traditional tools like newspapers, magazines, television shows, and billboards had been unable to tap. The smartphone helped companies like Google and Facebook storm these remaining redoubts of unmolested focus and start ransacking—generating massive new fortunes in the process.

Figuring out how to turn smartphones into ubiquitous billboards was not simple. As I mentioned in chapter 1, the original motivation for the iPhone was to prevent people from having to carry both an iPod and a cell phone in their pocket. To build a new sector of the economy on the back of this device required somehow convincing people to start looking at their phone . . . *a lot.* It was this directive that led companies like Facebook to innovate the field of attention engineering, figuring out how to exploit psychological vulnerabilities to trick users into spending far more time on these services than they actually intended. The average user now spends fifty minutes *per day* on Facebook products alone. Throw in other popular social media services and sites, and this number grows much larger. This type of compulsive use is not an accident, it's instead a fundamental play in the digital attention economy playbook.

To sustain this type of compulsive use, however, you cannot have people thinking too critically about how they use their phone. With this in mind, Facebook has in recent years presented itself as a *foundational technology*, like electricity or mobile telephony—something that everyone should just use, as it would be weird if you didn't. This status of cultural ubiquity is ideal for Facebook because it pressures people to remain users without having to sell them on concrete benefits.* An atmosphere of vagueness leads people to sign into the service with no particular purpose in mind, which, of course, makes them easier targets for the attention engineers' clever hooks and exploits—leading to the staggering amounts of usage time that Facebook needs to sustain its equally staggering $500 billion valuation.

Which brings us back to Facebook's folly. The reason why Ginsberg and Burke's article should concern their employer is because it punctures the myth of Facebook as a foundational technology that everyone should just "use" in some generic sense. By assessing different ways to engage with Facebook, one by one, identifying which seem more positive than

---

* As one of the rare Millennials who has never used Facebook, I have observed the reality of this vague cultural pressure through personal experience. As I've mentioned elsewhere, by far one of the most common arguments I used to hear from people about why I should sign up for Facebook is that there might be *some* benefit I didn't even know about that I might be missing. "You never know, maybe you'll find this to be useful" has got to be one of the worst product pitches ever devised. But in the peculiar context of the digital attention economy, it makes a lot of sense to people.

others, Ginsberg and Burke are encouraging people to think critically about what exactly they want to get out of this service.

This mind-set is potentially disastrous for the company. To see why, try the following experiment. Assuming that you use Facebook, list the most important things it provides you—the particular activities that you would really miss if you were forced to stop using the service altogether. Now imagine that Facebook started charging you by the minute. How much time would you really need to spend in the typical week to keep up with your list of important Facebook activities? For most people, the answer is surprisingly small; somewhere around twenty to thirty minutes.

The average Facebook user, by contrast, spends around 350 minutes per week on this company's services (if we take the fifty minutes per day cited above and multiply it by the seven days in a week). This means that if you were careful, you would be using these services somewhere around eleven to seventeen times less than average. If everyone started thinking about their use in similarly utilitarian terms—the terms promoted by Ginsberg and Burke—the amount of eyeball minutes Facebook has available to sell to advertisers would drop by more than an order of magnitude, creating a massive hit to their bottom line. Investors would revolt (in recent years, even single-digit reductions to Facebook quarterly earnings have stoked Wall Street anxiety), and the company would likely not survive with anything near its current form. Critical use is a critical problem for the digital attention economy.

■ ■ ■

Understanding the fragile attention economics that support companies like Facebook helps reveal an important strategy for succeeding with digital minimalism. The Ginsberg and Burke article highlights two radically different ways to think about *using* a service like Facebook. The big companies want "use" to be a simple binary condition—either you engage with their foundational technology, or you're a weirdo. By contrast, the type of "use" these companies perhaps most fear is the Ginsberg and Burke definition, which sees these products as offering a variety of different free services that you can carefully sift through and use in a manner that optimizes the value you receive.

This latter type of "use" is pure digital minimalism, but it's also difficult to successfully put into action. One of the reasons I detailed the specific financial numbers involved in the digital attention economy is to emphasize the sheer volume of resources these companies can deploy to push you away from the targeted use of Ginsberg and Burke, and toward the more open-ended wandering their business model depends on.

The lopsidedness of this battle is a big part of the reason I never messed around with any of these services in the first place. To repeat a line from the *New Yorker* writer George Packer, "[Twitter] scares me, not because I'm morally superior to it, but because I don't think I could handle it. I'm afraid I'd end up letting my son go hungry." If you must use these services, however, and you hope to do so without ceding autonomy over your time and attention, it's crucial to understand that this is not a casual decision. You're instead waging a David and Goliath battle against institutions that are both impossibly rich and intent on using this wealth to stop you from winning.

Put another way, to approach attention economy services with the intentionality proposed by Ginsberg and Burke is not a commonsense adjustment to your digital habits, but is instead better understood as a bold act of resistance. Fortunately, if you take this path, you'll not be alone. My research on digital minimalism has revealed the existence of a loosely organized *attention resistance* movement, made up of individuals who combine high-tech tools with disciplined operating procedures to conduct surgical strikes on popular attention economy services—dropping in to extract value, and then slipping away before the attention traps set by these companies can spring shut.

The remainder of this chapter, which is entirely dedicated to concrete advice, will bring you inside the tactics innovated by this resistance movement. The practices that follow each focus on a different category of these tactics. All of them have proved successful in shunting aside relentless efforts to capture your attention.

Perhaps more important than the details of these practices is the mind-set they embody. If your personal brand of digital minimalism requires engagement with services like social media, or breaking news sites, it's important to approach these activities with a sense of zero-sum antagonism. You want something valuable from their networks, and they want to undermine your autonomy—to come out on the winning side of this battle requires both preparation and a ruthless commitment to avoiding exploitation.

Vive la résistance!

PRACTICE: DELETE SOCIAL MEDIA FROM YOUR PHONE

Something big happened to Facebook starting around 2012. In March of that year, they began, for the first time, to show ads on the mobile version of their service. By October, 14 percent of the company's ad revenue came from mobile ads, making it into a small but nicely profitable piece of Mark Zuckerberg's growing empire. Then it took off. By the spring of 2014, Facebook reported that 62 percent of its revenue came from mobile, leading the technology website *The Verge* to declare: "Facebook is a mobile company now." This statement has continued to prove accurate: by 2017, mobile ad revenue rose to 88 percent of their earnings, and is still climbing.

These Facebook statistics underscore a trend true of social media more generally: mobile pays the bills. This reality has important implications for the attention resistance. It emphasizes that the smartphone versions of these services are much more adept at hijacking your attention than the versions ac-

cessed through a web browser on your laptop or desktop computer. This difference is due in part to the ubiquitous nature
of smartphones. Because you always have the phone with you,
every occasion becomes an opportunity to check your feeds.
Before the mobile revolution, services like Facebook could
only monetize your attention during periods when you happened to be sitting at your computer.

There's also, however, a more ominous feedback loop at
play. As more people began to access social media services on
their smartphones, the attention engineers at these companies
invested more resources into making their mobile apps stickier. As discussed in the first part of this book, some of these
engineers' most ingenious attention traps—including the slot
machine action of swiping down to refresh a feed, or alarm-
red notification badges—are mobile-only "innovations."

Pulling together these pieces of evidence points to a clear
conclusion: if you're going to use social media, stay far away
from the mobile versions of these services, as these pose a significantly bigger risk to your time and attention. This practice, in other words, suggests that you remove all social media
apps from your phone. You don't have to quit these services;
you just have to quit accessing them on the go.

■ ■ ■

This strategy is classic digital minimalism. By removing your
ability to access social media at any moment, you reduce its ability to become a crutch deployed to distract you from bigger
voids in your life. At the same time, you're not necessarily

abandoning these services. By allowing yourself access (albeit less convenient) through a web browser, you preserve your ability to use specific features that you identify as important to your life—but on your own terms.

I started informally offering this advice soon after my last book, *Deep Work*, was released in early 2016. At the time, a lot of readers were nervous about my minimalist suggestion to quit social media services that didn't provide more benefits than harms. Accordingly, I began to suggest that they take the apps off their phones as a first step. Two things struck me about the feedback that began to trickle in. First, a nontrivial percentage of people who deleted the apps discovered that they essentially stopped using social media altogether. Even the small extra barrier of needing to log in to a computer was enough to prevent them from making the effort—revealing, often to their admitted surprise, that services they claimed were indispensable were in reality providing nothing more than convenient hits of distraction.

The second thing I noticed was that for people who did continue to use social media on their computers, their relationship to these services transformed. They began to sign in for specific, high-value purposes, and only do so every once in a while. Facebook use, for example, dropped down toward one or two checks a week for many of my readers who took the app off of their phone. For them, social media became one tool among many they sometimes use, and stopped acting as an omnipresent drain on their attention.

For these reasons, this advice likely frightens social media companies. They're happy to argue about the importance of their services or give examples of the good things they have provided society. But the one thing they definitely don't want you to notice is that the only really good reason to be accessing these services *on your phone* is to ensure companies like Facebook continue to enjoy steady quarterly growth.

PRACTICE: TURN YOUR DEVICES INTO SINGLE-PURPOSE COMPUTERS

In 2008, Fred Stutzman was a graduate student at the University of North Carolina, working on a doctoral dissertation about the role of new tools like social media to aid life transitions, such as heading off to college. Perhaps ironically, given the topic of his research, Stutzman struggled with this work because his internet-connected laptop offered too many enticing distractions. His solution was to start writing at a nearby coffee shop. This plan worked well until the building next to the coffee shop got Wi-Fi. Frustrated by his inability to escape the attractions of the internet, Stutzman programmed his own tool to block the network connections on his computer for set amounts of time. He called it, appropriately enough, Freedom.

Stutzman posted the tool online, where it soon began to gather a cult following. Realizing that he was onto something, he shelved his academic career to focus on the software full time. In the years that followed, the tool became more sophisticated. Instead of simply deactivating the internet, you can now use it to block custom lists of distracting websites and applications, and set up regular schedules that activate this blocking automatically. It also works across all of your devices, allowing a single click from your Freedom dashboard to activate blocking across your computers, phones, and tablets.

The tool has since been adopted by over 500,000 users, including, notably, the novelist Zadie Smith, who thanked Freedom by name in the acknowledgments of her critically acclaimed 2012 bestseller, *NW*, crediting the software for "creating the time" needed for her to finish the manuscript. Smith is not alone. Freedom's internal research reveals that its users gain, on average, 2.5 hours of productive time per day.

Despite the effectiveness of Freedom—and other similarly popular blocking tools such as SelfControl—its role in human computer interaction is often misunderstood. Consider, for example, the following quote from a profile of Stutzman that appeared in *Science*: "There's an even deeper irony, and also a retro element, in the idea of taking a powerful productivity machine like a modern laptop computer and shutting down some of its core functions in order to increase productivity."

This sentiment, that temporarily blocking features of a general-purpose computer reduces its potential, is common for skeptics of tools like Freedom. It's also flawed: it represents a misunderstanding of computation and productivity that benefits the large digital attention economy conglomerates much more than the individual users that they exploit.

■ ■ ■

To understand my above claim, some brief history is needed. Electromechanical machines that performed useful tasks were around before electronic computers. Many people forget, for example, that IBM was selling automatic tabulating machines to the US Census Bureau as early as the 1890s. Part of what made computers so revolutionary was that they were *general purpose*—the same machine could be programmed to perform many different tasks. This approach was a huge improvement over constructing separate machines for each computational application, which is why computing technology ended up transforming the twentieth-century economy.

The personal computer revolution that began in the 1980s carried this message of general-purpose productivity to individuals. An early print ad for the Apple II, for example, tells the story of a California store owner who uses his computer during the week to chart sales, then, during the weekends, totes it home to work on family finances with his wife. The idea that one machine could perform many different tasks was a key selling point.

It's this mind-set, that "general purpose" equals "productivity," that leads people to cast a skeptical eye on tools like Freedom that remove options from your computing experience. The problem with this mind-set, however, is that it jumbles the role of time in this type of productivity. What makes general-purpose computing powerful is that you don't need separate devices for separate uses, *not* that it allows you to do multiple things at the same time. The California store owner from the earlier Apple ad used his computer to chart sales during the week and balance his checkbook on the weekends. He wasn't trying to do both simultaneously.

Until recently in the history of electronic computing, there was no reason to make this distinction, as personal computers could run only one user-facing program at a time, and there was a high cost for the user to switch from one application to another, often involving floppy disks and arcane commands. Today, of course, this has changed. As Stutzman learned while trying to write his doctoral dissertation, jumping from a word processor to a web browser requires only a single quick click. As many have discovered, the rapid switching between different applications tends to make the human's interaction with the computer less productive in terms of the quality and quantity of what is produced.

With this in mind, there's nothing deeply ironic about "taking a powerful productivity machine like a modern laptop computer and shutting down some of its core functions in order to increase productivity." It's instead quite natural once you recognize that the power of a general-purpose computer is in the total number of things it enables the user to do, not the total number of things it enables the user to do simultaneously.

As I hinted earlier, a major beneficiary of the reluctance to shut down features on your computer is the digital attention economy. When you allow yourself, at all points, access to all that your general-purpose computers can offer, this list will include apps and websites engineered to hijack your attention.

If you want to join the attention resistance, one of the most important things you can therefore do is follow Fred Stutzman's lead and transform your devices—laptops, tablets, phones—into computers that are general purpose in the long run, but are effectively single purpose in any given moment. This practice suggests that you use tools like Freedom to aggressively control when you allow yourself access to any website or app supported by a company that profits from your attention. I'm not talking about occasionally blocking some sites when working on a particularly hard project. I want you instead to think about these services as being *blocked by default*, and made available to you on an intentional schedule.

If you don't need social media for your work, for example, set up a schedule that blocks these sites and apps completely with the exception of a few hours in the evening. If you do need a particular social media tool for work (say, Twitter), then put aside a few blocks during the day when you can check it, and leave it otherwise blocked. If there are certain infotainment sites that pull at your attention (for me, for example, baseball news about the Washington Nationals becomes impossibly appealing at times), follow this habit of leaving these sites blocked by default outside of specific windows.

This practice of default blocking might at first seem overly aggressive, but what it's actually doing is bringing you back closer to the ideal of single-purpose computing that's much more compatible with our human attention systems. As with all of the advice in this chapter on the attention resistance, default blocking doesn't require you to abstain completely from the fruits of the digital attention economy but forces you to approach them with more intention. It's a different way of thinking about your relationship with your computer, and one that is becoming increasingly necessary to remain a minimalist in our current age of distraction.

PRACTICE: USE SOCIAL MEDIA LIKE A PROFESSIONAL

Jennifer Grygiel is a social media pro. I don't mean this in the colloquial sense that they (Jennifer prefers the pronoun "they/their" to "she/her") are good at using social media. I mean instead that Jennifer makes a living from an expert understanding of how to extract maximum value from these tools.

During the rise of the Web 2.0 revolution, Jennifer was the social business and emerging media manager at State Street, a global financial services firm headquartered in Boston. Jennifer helped the company build an internal social network that enabled employees around the world to collaborate more efficiently, and established State Street's social listening program—allowing them to more carefully monitor references to "State Street" amid the noise of typical social media chatter (a task, Jennifer told me, that's made particularly challenging when your company's name is found on thousands of road signs across the country).

From State Street, Jennifer moved to academia to become

an assistant professor of communication, specializing in social media, at the prestigious S.I. Newhouse School of Public Communications at Syracuse University. Jennifer now teaches a new generation of communication professionals how to maximize the power of social media.

As you might expect, given this career history, Jennifer spends a fair amount of time using social media. What interests me more than the total amount of time that Jennifer spends on social media is the details of *how* they use it. If you ask Jennifer about these habits, as I did while researching this chapter, you'll discover that social media professionals like Jennifer approach these tools differently than the average user. They seek to extract large amounts of value for their professional and (to a lesser extent) personal lives, while avoiding much of the low-value distraction these services deploy to lure users into compulsive behaviors. Their disciplined professionalism, in other words, provides a great example for any digital minimalist looking to join the attention resistance.

With this in mind, the remainder of this practice describes Jennifer's social media habits. You don't have to exactly mimic this particular mix of strategies, but this practice asks that you consider applying a similar level of intention and structure to your own engagement with these services.

■ ■ ■

In summarizing Jennifer Grygiel's social media habits, it's perhaps easiest to start with what Jennifer does *not* do. For one thing, Jennifer does not see social media as a particularly good source of entertainment: "If you [look at my Twitter feed,] you won't see a lot of dog meme accounts. . . . I already seem to get a lot of dog memes without needing to follow those accounts."

Jennifer does use Instagram to follow accounts from a small number of communities related to their interests—a sufficiently narrow focus that it typically takes only a few minutes to browse all new posts since the last check. Jennifer is more suspicious, however, of the increasingly popular Instagram Stories feature, which lets you broadcast moments of your life. Jennifer describes it as "reality TV starring your friends." This feature was introduced to increase the amount of content users generate, and therefore the amount of time they spend consuming this content. Jennifer's not biting: "I don't know if there's too much value added in that [feature]."

Jennifer also uses Facebook significantly less than the average user by maintaining a simple rule regarding the service: it's only for close friends and relatives, and for occasionally connecting with influencers. "In the early years, I used to accept friend requests from anyone," they said. "But I don't think we're really supposed to be connected to so many people so frequently." Jennifer now tries to keep friend engagement*

---

* Jennifer still has over 1,000 contacts on Facebook (it is a difficult social act to formally "unfriend" someone) but tries to limit active engagement to a count below the Dunbar Number. Jennifer uses the See First feature on their newsfeed and constraints on who gets messaged to help accomplish this engagement goal.

below the Dunbar Number of 150—a theoretical limit for the number of people a human can successfully keep track of in their social circles. Jennifer does not, for the most part, interact with professional colleagues on Facebook: "If I need to connect with a colleague, I'll stop by their office or chat after work." Jennifer also thinks it's not the right platform to keep up with news (more soon on what Jennifer prefers for this purpose) or to debate issues, noting "the civility issues on that platform have gotten difficult."

Instead, Jennifer logs on to Facebook maybe once every four days or so to see what's going on with their close friends and relatives. And that's it. The average user spends thirty-five minutes per day on Facebook's core functions (an amount that expands to around fifty minutes when you include the other social media services that Facebook owns). Jennifer typically spends less than an hour per week on the service. Checking in on your close social circles is a useful feature, but it's not one that requires a lot of time (a reality Facebook hopes you ignore).

Where Jennifer dedicates most of their social media attention these days is Twitter, which they believe, at this current moment, to be the most important service for professionals. Jennifer's reasoning for this belief is that in most fields, many prominent people tweet. By tapping into their collective wisdom, you can stay up to speed on breaking news and novel ideas. Twitter also exposes you to people who might be valuable to add to your professional network. (On many occasions during their career, Jennifer has benefited by reaching out through email to individuals that they've discovered through social media.*)

Drawing on their experience developing corporate social listening programs, Jennifer recognizes the overwhelming noisiness of most social media streams, and the care and discipline required to find useful signals in this noise. With this in mind, Jennifer maintains separate Twitter accounts for their academic interests and side interest in music (Jennifer played in bands for years). Within each account, Jennifer invests significant effort in selecting who they follow—focusing on high-quality thinkers, or similar influencers in their topic area. In their academic account, for example, Jennifer follows a curated list of journalists, technologists, academics, and policy makers.

Jennifer deploys Twitter as an early detection radar for trending news or ideas. This is particularly important for Jennifer's job, as they are often asked to give quotes or react to breaking news in their areas of expertise. When something catches Jennifer's attention on a social media timeline, they'll isolate it and dive deeper. In some cases, Jennifer will deploy a desktop tool called TweetDeck to aid this process. TweetDeck allows them to perform sophisticated searches to better understand Twitter trends. One important search function pro-

---

* This is how Jennifer and I first connected: Jennifer encountered my book through a recommendation, and then used social media to research my background, which uncovered the fact that we had come close to overlapping at MIT. Jennifer emailed me based on this foundation—sparking a friendly ongoing conversation about social media.

vided by this tool, for example, is *thresholding.* Here's how Jennifer explains it:

> I can search for a certain topic, say Black Lives Matter, and then set a threshold in TweetDeck that allows me to listen to this topic, but only see tweets with 50 likes or retweets. I can then refine this and say just show me the verified accounts.

Thresholding is just one type of advanced search allowed by TweetDeck, and TweetDeck is just one tool among many that allow this style of more advanced filtering (for this purpose, big companies often rely on expensive software suites that integrate with their customer relationship management systems). The more important takeaway message here is the sophistication with which pros like Jennifer cut through the noise of social media to identify what information regarding a trend is worth their attention.

■ ■ ■

"There's real opportunity in social media to really benefit and grow, and some real negative sides to it as well," Jennifer told me. "It's really like a tightrope . . . most of us need to find a balance." Professionals like Jennifer highlight an effective way of achieving this balance: approach social media as if you're the director of emerging media for your own life. Have a careful plan for how you use the different platforms, with the goal of "maximizing good information and cutting out the waste."

To a social media pro, the idea of endlessly surfing your feed in search of entertainment is a trap (these platforms have been designed to take more and more of your attention)—an act of being used by these services instead of using them to your own advantage. If you internalize some of this attitude, your relationship with social media will become less tempestuous and more beneficial.

PRACTICE: EMBRACE SLOW MEDIA

Early in 2010, a trio of Germans with backgrounds in sociology, technology, and market research posted online a document titled "Das Slow Media Manifest." The English translation reads: "The Slow Media Manifesto."

The manifesto opens by noting that the first decade of the twenty-first century "brought profound changes to the technological foundations of the media landscape." The second decade, the manifesto then proposes, should be dedicated to figuring out the "appropriate reaction" to these massive changes. Its suggestion: embrace the concept of "slow." Following the lead of the Slow Food movement—which promotes local food and traditional cuisine as an alternative to fast food, and which has become a major cultural force in Europe since its inception in Rome in the 1980s—the Slow Media Manifesto argues that in an age in which the digital attention economy is shoveling more and more clickbait toward us and fragmenting our focus into emotionally charged shards, the right response is to become more *mindful* in our media consumption:

> Slow Media cannot be consumed casually, but provoke the full concentration of their users. . . . Slow Media measure themselves in production, appearance and content against high standard of quality and stand out from their fast-paced and short-lived counterparts.

This movement remains predominantly European. In the United States, by contrast, our response to these same issues has proved more puritanical. Whereas the Europeans suggest transforming the consumption of media into a high-quality experience (much like the Slow Food movement approach to eating), Americans tend to embrace the "low information diet": a concept first popularized by Tim Ferriss, in which you aggressively eliminate sources of news and information to help reclaim more time for other pursuits. This American approach to information is much like our approach to healthy eating, which focuses more on aggressively eliminating what's bad than celebrating what's good.

There are merits to both approaches, but when it comes to navigating news and related information without becoming a slave to the attention economy conglomerates, I suspect the European focus on slowness is more likely to succeed in the long run. Embracing the Slow Media movement, therefore, is exactly what this practice suggests.

■ ■ ■

The original Slow Media Manifesto addresses both producers and consumers of media. I want to focus here just on consumption, with a particular emphasis on the news—as this is an aspect of media consumption that makes us particularly vulnerable to attention exploitation.

Many people now consume news by cycling through a set sequence of websites and social media feeds. If you're interested in politics, for example, and lean toward the left side of the political spectrum, this sequence might go from CNN .com, to the *New York Times* homepage, to *Politico*, to the *Atlantic*, to your Twitter feed, and finally to your Facebook timeline. If you're into technology, Hacker News and Reddit might be in that list. If you're into sports, you'll include ESPN.com and team-specific fan pages, and so on.

Crucial to this news consumption habit is the ritualistic nature of the sequence. You don't make a conscious decision about each of the sites and feeds you end up visiting; instead, once the sequence is activated, it unfolds on autopilot. The slightest hint of boredom becomes a trip wire to activate this whole hulking Rube Goldberg apparatus.

We're used to this behavior, so it's easy to forget that it's largely an artifact of the recent rise of the digital attention economy. These companies love your ritualistic checking, as each pass through your personal cycle deposits some more pennies in their bank account. Checking ten different sites ten times a day makes them money, even if it doesn't leave you more informed than checking one good site once a day. This behavior, in other words, is not a natural reaction to an increasingly connected age, but instead a lucrative tic bolstered by powerful economic pressures.

Slow Media offers a more palatable alternative.

To embrace news media from a mind-set of slowness requires first and foremost that you focus only on the highest-quality sources. Breaking news, for example, is almost always much lower quality than the reporting that's possible once an event has occurred and journalists have had time to process it. A well-known journalist recently told me that following a breaking story on Twitter gives him the *sense* that he's receiving lots of information, but that in his experience, waiting until the next morning to read the article about the story in the *Washington Post* almost always leaves him more informed. Unless you're a breaking news reporter, it's usually counterproductive to expose yourself to the fire hose of incomplete, redundant, and often contradictory information that spews through the internet in response to noteworthy events. Vetted reporting appearing in established newspapers and online magazines tends to provide more quality than social media chatter and breaking-news sites.

Similarly, consider limiting your attention to the best of the best when it comes to selecting individual writers you follow. The internet is a democratizing platform in the sense that anyone can share their thoughts. This is laudable. But when it comes to reporting and commentary, you should constrain your attention to the small number of people who have proved to be world class on the topics you care about. This doesn't necessarily mean that they have to write for a big established organization—a powerful voice expressing herself on a personal blog can be just as high quality as a longtime reporter for the *Economist*—but instead that they've proved to you to be reliably smart and insightful with their writing. When an issue catches your attention, in other words, you're usually better served checking in on what the people you respect most think about it than wading into the murk of a Twitter hashtag search or the back-and-forth commenting littering your Facebook timeline. It's a general rule of slow movements that a small amount of high-quality offerings is usually superior to a larger amount of low-quality fare.

Another tenet of slow news consumption: if you're interested in commentary on political and cultural issues, this experience is almost always enhanced by also seeking out the best arguments *against* your preferred position. I live in Washington, DC, so I know professional political operatives on both sides of the aisle. A requirement of their job is that they keep up to speed on the best opposing arguments. A side effect of this requirement is that they tend to be much more interesting to chat with about politics. In private, they don't exhibit the same anxious urge to tilt at straw man versions of opposing viewpoints that's exhibited by most amateur political commenters, and instead are able to isolate the key underlying issues, or identify the interesting nuances that complicate the matter at hand. I suspect they derive much more pleasure out of consuming political commentary than those who merely seek confirmation that anyone who disagrees is deranged. As we've known since the time of Socrates, engaging with arguments provides a deep source of satisfaction independent of the actual content of the debate.

Another important aspect of slow news consumption is the decisions you make regarding *how* and *when* this consumption occurs. The compulsive click cycle described earlier is the news equivalent of snacking on Doritos, and is not compatible with the principles of the slow movement. I recommend instead isolating your news consumption to set times during the week. To foster the state of "full concentration" promoted by the Slow Media Manifesto, I further recommend that you ritualize this consumption by choosing a location that will support you in giving your full attention to the reading. I also recommend that you care about the particular format in which you do this reading.

For example, perhaps you look through an old-fashioned paper newspaper each morning over breakfast. This brings you up to speed on the major stories and provides a more interesting mix of stories than what you would curate for yourself online. Then, on Saturday mornings, perhaps you check in on a carefully selected group of online news sites and columnists, bookmarking the articles you want to dive deeper into, before heading to a local coffee shop with your tablet to read through this week's worth of deeper articles and commentary. If you can download these articles in advance, allowing you to read them without the distractions offered by an internet connection, that's even better. Serious news consumers also tend to deploy browser plug-ins or aggregation tools that can present them with articles stripped clean of advertisements and clickbait.

If you follow the above approach to news consumption (or something with a similar focus on slowness and quality), you will remain informed about current events and up to speed on big ideas in the spaces you care most about. But you will also accomplish this without sacrificing your time and emotional health to the frantic cycle of clicking that defines so many people's experience of the news.

There are any number of other rules and rituals that can offer similar benefits. The key to embracing Slow Media is the general commitment to maximizing the quality of what you consume and the conditions under which you consume it. If you're serious about joining the attention resistance, you should be serious about these ideas when confronting how you interact with information on the internet.

PRACTICE: DUMB DOWN YOUR SMARTPHONE

Paul works for a midsize industrial company in the United Kingdom. He's not a senior citizen. In fact, he's relatively young. I'm telling you this to underscore the unusual step Paul took in the fall of 2015, when he traded in his smartphone for a Doro PhoneEasy—a basic clamshell flip phone with oversize buttons and big-font display, marketed mainly to elderly people.*

I asked Paul about the experience. "It's silly, I know, but the first few weeks felt rough," he told me. "I didn't know what to do with myself." But then came the benefits. One of the major positive changes was that he no longer felt like his attention was divided when he was with his wife and kids. "I hadn't appreciated how distracted I had been around them." While at work, his productivity shot up. Meanwhile, after those rough initial few weeks, he felt the sense of boredom and jitteriness dissipate. "I feel less anxious. I hadn't realized how anxious I'd become." His wife told him that she was struck by how happy he now seems.

When the technology executive Daniel Clough decided to dumb down his phone experience, he didn't trash his iPhone but instead put it in the kitchen cupboard. He likes to use it when exercising so that he can listen to music and run his Nike+ fitness tracking app. On most other occasions, however, he brings his Nokia 130, a sleeker version of the Doro that shares its simplicity: no camera, no apps, no web—only calls and text messages. Like Paul, it took Clough a week or so to

---

* Interestingly, Paul later discovered that there's an underground movement of executives that use dumb phones like the Doro. They are, for the most part, in finance—typically hedge fund managers. It turns out that for people who move hundreds of millions of dollars in high-stakes trades every day, there's great advantage in shielding yourself from distracting market information that can bias your decisions and potentially cost you massive amounts of money.

overcome the urge to constantly check something, but he soon passed that hurdle. As he reports on his personal blog: "I feel so much better. I'm more present and my mind feels less cluttered." According to Clough, the main inconvenience he experiences about life without a smartphone is his inability to google something on the go: "But how great I feel without a smartphone far outweighs that."

Even *The Verge*, a reliable bastion of techno-boosterism, admitted the potential value of a return to simpler communication devices. Exhausted by the near constant Twitter checking induced by the 2016 presidential election, reporter Vlad Savov wrote an article titled "It's Time to Bring Back the Dumb Phone," in which he claims that a return to simpler phones "is not as drastic a regression as you might think—or as it might have been a few years ago." His main argument is that tablets and laptops have become so lightweight and portable that there is no longer a need to try to cram productivity functionalities into increasingly powerful (and therefore increasingly distracting) smartphones—phones can be used for calls and messages, and other portable devices can be used for everything else.

Some people want both options—the ability to take a smartphone with them on some occasions (longer trips, or when they might need to use a particular app), and a nondistracting simpler device on other occasions—but worry about the inconvenience of maintaining two different numbers. There's now a solution for this scenario as well: the tethered dumb phone. These products, which include, notably, a

Kickstarter darling called the Light Phone, don't replace your existing smartphone, but instead extend it to a simpler form.

Here's how it works. Let's say you have a Light Phone, which is an elegant slab of white plastic about the size of two or three stacked credit cards. This phone has a keypad and a small number display. And that's it. All it can do is receive and make telephone calls—about as far as you can get from a modern smartphone while still technically counting as a communication device.

Assume you're leaving the house to run some errands, and you want freedom from constant attacks on your attention. You activate your Light Phone through a few taps on your normal smartphone. At this point, any calls to your normal phone number will be forwarded to your Light Phone. If you call someone from it, the call will show up as coming from your normal smartphone number as well. When you're ready to put the Light Phone away, a few more taps turns off the forwarding. This is not a replacement for your smartphone, but instead an escape hatch that allows you to take long breaks from it.

The creators of the Light Phone, Joe Hollier and Kaiwei Tang, met inside a Google incubator, where they were encouraged to make software apps and taught about what makes these products desirable to funders. They were not impressed. "Quickly it became obvious that the last thing the world needed was another app," they write on their website. "Light was born as an alternative to the tech monopolies that are fighting more and more aggressively for our attention." Just in case their intentions as members of the attention resistance were not clear enough, Hollier and Tang posted a manifesto that opens with a diagram that reads: "Your [clock symbol] = Their [money symbol]."

■ ■ ■

In my earlier chapter on solitude, I suggested that you reject the mind-set that says you must *always* have your smartphone with you. The hope was to create more occasions for solitude— which we as humans need to thrive. The examples discussed here go much further, as they suggest the possibility of acquiring an alternative communication device that allows you to spend most (if not all) of your time free from a smartphone.

Declaring freedom from your smartphone is probably the most serious step you can take toward embracing the attention resistance. This follows because smartphones are the preferred Trojan horse of the digital attention economy. As discussed at the opening of this chapter, it was the spread of these always-on, interactive billboards that allowed this niche sector to expand to the point that they now enjoy as dominant players in the worldwide economy. Given this reality, if you're not carrying a smartphone, you fall off the radar of these organizations, and as a result, you'll find your efforts to reclaim your attention significantly simplified.

Dumbing down your phone, of course, is a big decision. Our attraction to these devices goes well beyond their ability to provide distraction. For many, they provide a safety net for modern life—protection against being lost, feeling alone, or missing out on something better. Convincing yourself that a dumb phone can satisfy enough of these needs so that its benefits outweigh its costs is not necessarily easy. Indeed, it might require a leap of faith—a commitment to test life without a smartphone to see what it's really like.

For others, this practice may remain too extreme. Some people are tied to their smartphones for specific reasons that cannot be ignored. If you're a health care worker who makes home visits, for example, maintaining access to Google Maps is key. Similarly, around the time I was writing this chapter, I received a note from a reader from Curitiba, Brazil, noting that the ability to use ride-sharing services like Uber and 99 is crucial to getting around in a city where cabs and walking are often not available options.

For other people, the opposite issue might hold: their smartphones aren't *enough* of a problem for them to receive much benefit from removing them from their life. I count myself in this category. I don't have any social media accounts, I don't play mobile games, I'm terrible about texting, and I already spend long times away from my phone each day. I could turn in my used iPhone for a Nokia 130, but I don't think it would make much difference.

On the other hand, if you're someone who could conceivably get away without ubiquitous smartphone access, and if your gut is telling you that this might make your life much better, then you should be reassured that this decision is no longer as radical as it might have once seemed. The dumb phone movement is gathering steam, and the tools available to support this lifestyle change are improving. If you're exhausted by your smartphone addiction, it's not only possible to say, "No more," it's actually not that hard. Remember how Hollier and Tang opened their manifesto with the idea "Your Time = Their Money." You should feel empowered to instead invest this value in things that matter more to you.

Conclusion

I n the fall of 1832, a French packet ship named *Sully* left Le Havre en route to New York. On board was a forty-one-year-old painter traveling home from a European tour in which his work had failed to generate much notice. His name was Samuel Morse.

As the historian Simon Winchester recounts, it was on this journey, somewhere in the middle of the Atlantic, that Morse "experienced the epiphany that would help him change the world." The catalyst for this moment was a fellow passenger, Charles Jackson, a Harvard geologist who happened to be up to date on recent discoveries in the study of electricity. As the two men discussed potential uses for this new medium, they stumbled across a remarkable insight. As Morse recalls thinking: "If the presence of electricity can be made *visible* in any part of the circuit, I see no reason why intelligence may not be transmitted by electricity."

In Winchester's telling, this was a "vatic revelation" for the failed painter, who immediately understood the possibilities of electronic communication. On arrival in New York, he rushed to his studios to begin the long process of experimentation needed to make practical the deceptively simple idea hatched on the *Sully*. Twelve frenetic years later, in May of 1844, Morse set up his telegraph key on a table in the chamber of the United States Supreme Court, where he was surrounded by a small group of influential legislators and government officials. An electrical wire, augmented on regular intervals with signal-boosting relays, connected Morse to his associate and fellow inventor Alfred Vail, who was stationed forty miles away in a railway station outside of Baltimore.

It was time for Morse to make his first major demonstration of his invention. All he needed was an inaugural message. Based on a suggestion from the daughter of the patent commissioner who had supported Morse's innovation, he tapped a well-known phrase from the end of the book of Numbers: WHAT HATH GOD WROUGHT?

As Winchester notes, these words, when considered in isolation, "formed a simple declarative exclamation, a statement of Samuel Morse's faith." But in the context of the transformation this invention and its successors would spark, it was better understood as a "suitably portentous epigraph for an era of change that now commenced with unimagined speed and unimaginable consequences."

Humans have been improving their world with invention since before the beginning of recorded history. But there's something about the innovations driving electronic communication that make them, as Winchester writes, "so mystifyingly different from what had gone before." Mechanical wonders fit the physical understanding of the world etched into our brains through millions of years of evolution. A charging steam locomotive might be awe inspiring, but it also fundamentally makes sense: fire creates steam that pushes the train's pistons forward.

A telegraph message or phone call or email or social media ping is somehow different. We lack an intuition for flowing electricity and the complex components that control it, and the concept of back-and-forth conversation existing outside the context of two people talking in close proximity is completely foreign to our species' history. The result is that we've always struggled to imagine the consequences of the electronic communication revolution started by Samuel Morse, and often find ourselves scrambling to make ex post facto sense of its impacts on our world.

As noted in an earlier chapter, Henry David Thoreau's reaction to the telegraph boom that followed Morse's 1844 demonstration was to remark that we're so eager to build a line between Maine and Texas that we never stopped to ask why these two states needed to be connected in the first place. Though dated in its particulars, this same sentiment applies well to our current age of social media and smartphones. First Facebook, then the iPhone: compulsive communicating and connecting— supported by mysterious, almost magical innovations in radio modulation and fiber-optic routing—swept our culture before anyone had the presence of mind to step back and re-ask Thoreau's fundamental question: *To what end?*

The result is a society left reeling by unintended consequences. We eagerly signed up for what Silicon Valley was selling, but soon realized that in doing so we were accidently degrading our humanity.

It's in this long trajectory that we can place digital minimalism. This philosophy is meant to be a human bulwark against the foreign artificiality of electronic communication, a way to take advantage of the wonders that these innovations do in fact provide (Maine and Texas, it turns out, *did* have some useful things to say once connected), without allowing their mysterious nature to subvert our human urge to build a meaningful and satisfying life.

■ ■ ■

This history places digital minimalism in a somewhat grandiose position, but as we explored in the preceding chapters, *implementing* this philosophy is largely an exercise in pragmatism. Digital minimalists see new technologies as tools to be used to support things they deeply value—not as sources of value themselves. They don't accept the idea that offering *some* small benefit is justification for allowing an attention-gobbling service into their lives, and are instead interested in applying new technology in highly selective and intentional ways that yield big wins. Just as important: they're comfortable missing out on everything else.

At the same time, I want to emphasize that transitioning to this lifestyle can be demanding—many of the minimalists I interviewed balanced their tales of triumph with examples where they let a tool get the best of them. This is fine. Adopting digital minimalism is not a onetime process that completes the day after your digital declutter; it instead requires ongoing adjustments.

In my experience, the key to sustained success with this philosophy is accepting that it's not really about technology, but is instead more about the quality of your life. The more you experiment with the ideas and practices on the preceding pages, the more you'll come to realize that digital minimalism is much more than a set of rules, it's about cultivating a life worth living in our current age of alluring devices.

Those who are committed to the digital status quo might attempt to cast this philosophy as somehow anti-technology. I hope I've convinced you in this book that this claim is misguided. Digital minimalism definitively does not reject the innovations of the internet age, but instead rejects the way so many people currently engage with these tools. As a computer scientist, I make a living helping to advance the cutting edge of the digital world. Like many in my field, I'm enthralled by the possibilities of our techno-future. But I'm also convinced that we cannot unlock this potential until we put in the effort required to take control of our own digital lives—to confidently decide for ourselves what tools we want to use, for what reasons, and under what conditions. This isn't reactionary, it's common sense.

I opened this book with Andrew Sullivan's concern about losing his humanity in the electronic world wrought by Samuel Morse. "I used to be a human being," he wrote. My hope is that digital minimalism can help reverse this state of affairs by providing a constructive way to engage and leverage the latest innovations to *your* advantage, not that of faceless attention economy conglomerates, to create a culture where the technologically savvy can upend Sullivan's lament and instead say with confidence: "Because of technology, I'm a better human being than I ever was before."

Acknowledgments

The idea to write this book was born on a deserted beach, on an island in the Bahamas, during the last weeks of 2016. At the time, I was already well along in the process of researching a book on a completely different topic. But as I mentioned in the introduction, I had, by this point, begun hearing from readers of my last book, *Deep Work*, who were struggling with the role of new technologies in their personal lives. I couldn't shake the idea that this topic was too rich to ignore—the urgency with which people discussed it hinted that it was about much more than just smarter tech tips, but instead something that touches on the universal human aspiration to cultivate a good life.

With time to spare on vacation, and miles of empty beach to wander (we arrived before the busy season), I decided to dedicate some thought to a simple question: If I *were* to write a book on this topic, what would it be like? After a few days of contemplative wandering, a compelling phrase popped into my head: *digital minimalism.* From there, I began to furiously take notes, and an outline of a philosophy emerged.

My first step in validating this idea was to run it by my wife, Julie, who, in addition to being my best friend and an indefatigable mother to our three children, is my primary sounding board for all things related to my writing career. Her enthusiastic response motivated me to keep pushing forward. On returning home, I sent a note to my longtime literary agent and publishing world mentor, Laurie Abkemeier, floating the idea that we put my current project on hold to tackle this new idea first. She agreed and helped me immensely through the difficult process of shaping my loose ideas into a tightly focused book proposal, and then rolling out the proposal to the publishing world in such a way that they would share my excitement. I'm incredibly grateful for her tireless efforts during this demanding period.

I am, of course, also thankful to my editor, Niki Papadopoulos at Portfolio, as well as Adrian Zackheim, the founder and publisher of the imprint, for taking on this project and believing in its potential. Niki's guidance was invaluable in helping me transform the early drafts of my manuscript into something strong and compelling. I must also thank Vivian Roberson at Portfolio for her insightful help in polishing the manuscript and shepherding it through production, and Tara Gilbride for leading the publicity efforts for this book. Working with the entire team at Portfolio has been nothing but a pleasure. As an author, I couldn't have asked for a better experience.

Notes

INTRODUCTION

ix **"An endless bombardment of news"**: Andrew Sullivan, "I Used to Be a Human Being," *New York*, September 18, 2016, http://nymag.com/select all/2016/09/andrew-sullivan-my-distraction-sickness-and-yours.html.

xii **The techno-philosopher Jaron Lanier convincingly argues**: For more on Jaron Lanier's thoughts about the primacy of negativity in the attention marketplace, see his *Vox* podcast interview with Ezra Klein from January 16, 2018, https://www.vox.com/2018/1/16/16897738/jaron-lanier -interview.

xv **"simplicity, simplicity, simplicity"**: Henry David Thoreau, *Walden; or, Life in the Woods* (New York: Dover Publications, 2012), 59. Because the text of *Walden* is in the public domain, many different online, ebook, audio, and print editions of the book exist. I cite the Dover print edition for the purposes of providing page numbers. All quotes from *Walden* that I reference, however, correspond exactly to the public domain version of text (e.g., as accessible here: http://www.gutenberg.org/files/205/205-h/205-h.htm).

xv **"You see how few things"**: Marcus Aurelius, *Meditations*, trans. Gregory Hays (New York: Modern Library, 2003), 18.

xviii **"The mass of men lead lives"**: Thoreau, *Walden*, 4.

xviii **They honestly think**: Thoreau, *Walden*, 5.

CHAPTER 1: A LOPSIDED ARMS RACE

4 **"It's the best iPod we've ever made!"**: "Steve Jobs iPhone 2007 Presentation (HD)," YouTube video, 51:18, recorded January 9, 2007, posted

by Jonathan Turetta, May 13, 2013, https://www.youtube.com/watch?v=vN4U5FqrOdQ.

5 **"The killer app is making calls":** "Steve Jobs iPhone 2007."

5 **"This was supposed to be an iPod":** Andy Grignon, phone interview by the author, September 7, 2017.

6 **"a moment can feel":** Laurence Scott, *The Four-Dimensional Human: Ways of Being in the Digital World* (New York: W. W. Norton, 2016), xvi.

9 **The tycoons of social media:** "Social Media is the New Nicotine | Real Time with Bill Maher (HBO)," YouTube video, 4:54, posted May 12, 2017, https://www.youtube.com/watch?v=KDqoTDM7tio.

10 **"This thing is a slot machine":** Tristan Harris, interview with Anderson Cooper, *60 Minutes*, https://www.cbsnews.com/video/brain-hacking.

12 **"race to the bottom":** Bianca Bosker, "The Binge Breaker," *Atlantic*, November 2016, https://www.theatlantic.com/magazine/archive/2016/11/the-binge-breaker/501122.

12 **"serves us, not advertising":** This quote comes from an earlier version of the website for Time Well Spent. The organization has since been re-branded as the Center for Humane Technology, and has a new website and new copy: http://humanetech.com.

13 **Before 2013, Adam Alter had little interest in technology:** Adam Alter, phone interview by the author, August 23, 2017.

15 **"Addiction is a condition":** "Addiction," Substance Abuse, *Psychology Today*, https://www.psychologytoday.com/basics/addiction, accessed July 11, 2018.

16 **"growing evidence suggests that behavioral addictions":** Jon E. Grant, Marc N. Potenza, Aviv Weinstein, and David A. Gorelick, "Introduction to Behavioral Addictions," *American Journal of Drug and Alcohol Abuse* 36, no. 5 (2010): 233–41, https://www.ncbi.nlm.nih.gov/pmc/articles/PMC3164585.

17 **Scientists have known since:** Michael D. Zeiler and Aida E. Price, "Discrimination with Variable Interval and Continuous Reinforcement Schedules," *Psychonomic Science* 3, nos. 1–12 (1965): 299, https://doi.org/10.3758/BF03343147.

18 **"It's hard to exaggerate how much":** Adam Alter, *Irresistible: The Rise of Addictive Technology and the Business of Keeping Us Hooked* (Penguin Press, 2017), 128.

18 **"bright dings of pseudo-pleasure":** Paul Lewis, "'Our Minds Can Be Hijacked': The Tech Insiders Who Fear a Smartphone Dystopia," *Guardian*, October 6, 2017, https://www.theguardian.com/technology/2017/oct/05/smartphone-addiction-silicon-valley-dystopia.

19 **"Apps and websites sprinkle intermittent variable rewards":** Tristan Harris, "How Technology Is Hijacking Your Mind—from a Magician

and Google Design Ethicist," Thrive Global, May 18, 2016, https://me
dium.com/thrive-global/how-technology-hijacks-peoples-minds
-from-a-magician-and-google-s-design-ethicist-56d62ef5edf3.
19 **"but no one used it"**: Lewis, "'Our Minds Can Be Hijacked.'"
19 **"The thought process that went into building"**: Mike Allen, "Sean
Parker Unloads on Facebook: "God Only Knows What It's Doing to Our
Children's Brains," Axios, November 9, 2016, https://www.axios.com
/sean-parker-unloads-on-facebook-2508036343.html.
20 **"We're social beings"**: Alter, *Irresistible*, 217–18.
21 **"Whether there's a notification or not"**: Victor Luckerson, "The Rise
of the Like Economy," *The Ringer*, February 15, 2017, https://www
.theringer.com/2017/2/15/16038024/how-the-like-button-took-over
-the-internet-ebe778be2459.
22 **Tristan Harris highlights the example of tagging people in photos:**
Harris, "How Technology Is Hijacking."
23 **"It's a social-validation feedback loop"**: Allen, "Sean Parker Unloads."

CHAPTER 2: DIGITAL MINIMALISM
27 **"It's relatively easy to retake control"**: Leonid Bershidsky, "How I
Kicked the Smartphone Addiction—and You Can Too," *New York
Post*, September 2, 2017, http://nypost.com/2017/09/02/how-i-kicked-the
-smartphone-addiction-and-you-can-too.
29 **The so-called digital minimalists:** The case studies of digital
minimalists cited throughout this chapter come from email interactions
with the author.
36 **Henry David Thoreau borrowed an ax:** Thoreau, *Walden*, 26–27.
37 **"I went to the woods"**: Thoreau, *Walden*, 59.
38 **bland expense tables:** Thoreau, *Walden*, 39.
39 **philosopher Frédéric Gros calls:** Frédéric Gros, trans. John Howe, *A
Philosophy of Walking* (London: Verso, 2014), 90.
39 **"The cost of a thing"**: Thoreau, *Walden*, 19.
40 **"crushed and smothered under [their] load"**: Thoreau, *Walden*, 2.
40 **"mass of men lead[ing] lives"**: Thoreau, *Walden*, 4.
40 **"I see young men, my townsmen"**: Thoreau, *Walden*, 2.
42 **"The striking thing with Thoreau"**: Gros, *A Philosophy of Walking*, 90.
48 **"We need to reevaluate [our current relationship with]"**: Max Brooks,
interview by Bill Maher, *Real Time with Bill Maher*, HBO, November
17, 2017.
48–49 **"giv[ing] people the power to build community"**: "What Is Facebook's Mission Statement?," FAQs, Facebook Investor Relations, https://
investor.fb.com/resources/default.aspx, accessed July 11, 2018.

50 **"Amish communities are not relics"**: John A. Hostetler, *Amish Society*, 4th ed. (Baltimore: Johns Hopkins University Press, 1993), ix.

50 **"Amish lives are anything but antitechnological"**: Kevin Kelly, *What Technology Wants* (New York: Viking, 2010), 217.

50 **"cruising down the road"**: Kelly, *What Technology Wants*, 219.

50 **"smoking, noisy contraption"**: Kelly, *What Technology Wants*, 218.

51 **In one memorable passage**: Kelly, *What Technology Wants*, 221. Kelly is actually talking about a strict Mennonite family instead of an Amish family, but the border between strict Mennonites and normal Amish is blurred, so the example is relevant for our purposes.

51 **use technology: "Amish hacking"**: Jeff Brady, "Amish Community Not Anti-Technology, Just More Thoughtful," *All Things Considered*, NPR, September 2, 2013, https://www.npr.org/sections/alltechconsidered/2013 /09/02/217287028/amish-community-not-anti-technology-just-more -thoughtful.

51 **"Is this going to be helpful"**: Brady, "Amish Community Not Anti-Technology."

52 **"When cars first appeared"**: Kelly, *What Technology Wants*, 218.

52 **"When people leave the Amish"**: Donald B. Kraybill, Karen M. Johnson-Weiner, and Steven M. Nolt, *The Amish* (Baltimore: Johns Hopkins University Press, 2013), 325.

53 **"In any discussion about the merits"**: Kelly, *What Technology Wants*, 217.

54 **the percentage of Amish youth that decide to stay after Rumspringa**: "Rumspringa: Amish Teens Venture into Modern Vices," *Talk of the Nation*, NPR, June 7, 2006, https://www.npr.org/templates/story /story.php?storyId=5455572.

54 **The restrictions that guide each community, called the *Ordnung***: For more details on Amish society, including the operation of the *Ordnung* and the relative powerlessness of women, see the following survey by David Friedman: http://www.daviddfriedman.com/Academic/Course _Pages/legal_systems_very_different_12/Book_Draft/Systems/Amish Chapter.html.

55 **"I don't think I'd be a good smartphone user"**: Laura, phone interview by the author, December 16, 2017.

CHAPTER 3: THE DIGITAL DECLUTTER

61 **Our efforts even made national news**: Emily Cochrane, "A Call to Cut Back Online Addictions: Pitted Against Just One More Click," *New York Times*, February 4, 2018, https://www.nytimes.com/2018/02/04/us/politics /online-addictions-cut-back-screen-time.html.

63 **"restless without video games":** This quote, in addition to all other quotes in this chapter from digital declutter experiment participants, come from email correspondence with the author between December 2017 and February 2018.

66 **"feel more invested in the time I spend with people . . .":** Cochrane, "Call to Cut Back."

CHAPTER 4: SPEND TIME ALONE

87 **"This president had absolutely no honeymoon":** Henry Lee Miller, *President Lincoln: The Duty of a Statesman* (New York: Alfred A. Knopf, 2008), 48.

87 **"The first thing that was handed to me":** Miller, *President Lincoln*, 49. Miller cites the diary of Senator Browning as his source. For more detail, see *The Diary of Orville Hickman Browning*, vol. 1, ed. Theodore Calvin Pease and James G. Randall (Springfield: Illinois State Historical Library, 1925–33), 476.

88 **"Virtually from Lincoln's first day":** Harold Holzer, "Abraham Lincoln's White House," *White House History* 25 (Spring 2009), https://www.whitehousehistory.org/abraham-lincolns-white-house.

88 **The White House Historical Association preserves an engraving:** Holzer, "Abraham Lincoln's White House," see fifth photo.

88 **"the biggest drain on the president's time":** Holzer, "Abraham Lincoln's White House."

89 **"The servant who answered the bell":** John French quote from Matthew Pinsker, *Lincoln's Sanctuary: Abraham Lincoln and the Soldiers' Home* (New York: Oxford University Press, 2005), 52. This book provides a definitive modern history of Lincoln's time at the Soldiers' Home, and it is recommended to those looking to learn more on this topic.

90 **"was here at the cottage":** Erin Carlson Mast, interview with the author, October 6, 2017.

91 **The president would also famously:** For more on Lincoln's practice of recording ideas on scraps of paper, see Jeanine Cali, "Lincoln's Emancipation Proclamation—Pic of the Week," *In Custodia Legis: Law Librarians of Congress* (blog), Library of Congress, May 3, 2013, https://blogs.loc.gov/law/2013/05/lincolns-emancipation-proclamation-pic-of-the-week.

93 **"I get an extra 20 IQ points":** Raymond M. Kethledge, interview by David Lat, "Lead Yourself First: An Interview with Judge Raymond M. Kethledge," *Above the Law*, September 19, 2017, http://abovethelaw.com/2017/09/lead-yourself-first-an-interview-with-judge-raymond-m-kethledge/?rf=1.

93 **"running is cheaper than therapy"**: Raymond M. Kethledge and Michael S. Erwin, *Lead Yourself First: Inspiring Leadership through Solitude* (New York: Bloomsbury USA, 2017), 94.

94 **They note that King's involvement**: Kethledge and Erwin, *Lead Yourself First*, 155–56.

95 **"And it seemed at that moment"**: Kethledge and Erwin, *Lead Yourself First*, reproduces this quote on page 163; the primary source: Martin Luther King Jr., *Stride Toward Freedom: The Montgomery Story* (New York: Harper & Brothers, 1958).

95 **"the most important night"**: David Garrow, *Bearing the Cross* (New York: William Morrow, 1986; reprint, New York: William Morrow Paperbacks, 2004), 57.

95 **"All of humanity's problems"**: Blaise Pascal, *Pascal's Pensées*, Thought #139. This is one of the more common of the many English translations of this phrase.

96 **"I have read abundance of fine things"**: Benjamin Franklin, "Journal of a Voyage," August 25, 1726, Papers of Benjamin Franklin, digital edition, Yale University and Packard Humanities Institute, http://franklin papers.org/franklin/framedVolumes.jsp?vol=1&page=072a.

96 **"Conversation enriches the understanding"**: Anthony Storr, *Solitude: A Return to the Self* (1988; reprint, New York: Free Press, 2005), ix.

96 **"the majority of poets"**: Storr, *Solitude*, ix.

97 **"it matters enormously"**: Michael Harris, *Solitude: In Pursuit of a Singular Life in a Crowded World* (New York: Thomas Dunne Books, 2017), 40.

98 **"new ideas; an understanding"**: Harris, *Solitude*, 40.

98 **"the ability to be alone"**: Harris, *Solitude*, 39.

98 **"I am here alone"**: May Sarton, *Journal of a Solitude* (New York: W. W. Norton, 1992), 11. I first encountered this quote (with accompanying commentary) in Maria Popova, "May Sarton on the Cure for Despair and Solitude as the Seedbed for Self-Discovery," *Brain Pickings* (blog), October 17, 2016, https://www.brainpickings.org/2016/10/17/may-sarton-journal -of-a-solitude-depression.

98 **"We enter solitude"**: From the essay "Healing," Wendell Berry, *What Are People For?: Essays*, 2nd ed., (Berkeley: Counterpoint, 2010), 11.

99 **"contemporary Western culture"**: Storr, *Solitude*, 70.

99 **"we are in great haste"**: Thoreau, *Walden*, 34.

102 **Alter decided to measure his own smartphone use**: Alter, *Irresistible*, 13–14.

102 **"There are millions of smartphone users"**: Alter, *Irresistible*, 14.

103 **"Facebook . . . was built"**: "Facebook's Letter from Mark Zuckerberg— Full Text," *The Guardian*, https://www.theguardian.com/technology /2012/feb/01/facebook-letter-mark-zuckerberg-text

104 **The term *constant* is not hyperbole:** "Tweens, Teens, and Screens: What Our New Research Uncovers," Common Sense Media, November 2, 2015, https://www.commonsensemedia.org/blog/tweens-teens-and-screens-what-our-new-research-uncovers.

106 **"The gentle slopes of the line graphs":** Jean M. Twenge, "Have Smartphones Destroyed a Generation?," *The Atlantic*, September 2017, https://www.theatlantic.com/magazine/archive/2017/09/has-the-smartphone-destroyed-a-generation/534198.

106 **"Rates of teen depression and suicide":** Twenge, "Have Smartphones."

107 **"Much of this deterioration can be traced":** Twenge, "Have Smartphones."

107 **"Anxious kids certainly existed before Instagram":** Benoit Denizet-Lewis, "Why Are More American Teenagers Than Ever Suffering from Severe Anxiety?," *New York Times Magazine*, October 11, 2017, https://www.nytimes.com/2017/10/11/magazine/why-are-more-american-teenagers-than-ever-suffering-from-severe-anxiety.html.

107 **"To my surprise, anxious teenagers tended":** Denizet-Lewis, "American Teenagers."

107 **"It seemed like too easy an explanation":** Denizet-Lewis, "American Teenagers."

108 **"The use of social media":** Denizet-Lewis, "American Teenagers."

110 **"[Thoreau's] intention was not to inhabit":** W. Barksdale Maynard, "Emerson's 'Wyman Lot': Forgotten Context for Thoreau's House at Walden," *The Concord Saunterer: A Journal of Thoreau Studies*, no. 12/13 (2004/2005): 59–84, http://www.jstor.org/stable/23395273, quoted in Erin Blakemore, "The Myth of Henry David Thoreau's Isolation," JSTOR Daily, October 8, 2015, https://daily.jstor.org/myth-henry-david-thoreaus-isolation/.

111 **"I've always had a sort of intuition":** *Thirty Two Short Films about Glenn Gould*, directed by François Girard (Samuel Goldwyn Company, 1993), quoted in Harris, *Solitude*, 217.

112 **"We have zero tolerance for talking":** "About," Alamo Drafthouse Cinema, https://drafthouse.com/about, accessed July 14, 2018.

112 **"You can't tell a 22-year-old":** Adam Aron interview by Brent Lang, "AMC Executives Open to Allowing Texting in Some Theaters," *Variety*, April 13, 2016, http://variety.com/2016/film/news/amc-texting-theaters-phones-1201752978.

114 **A young woman named Hope King:** Hope King, "I Lived without a Cell Phone for 135 Days," CNN Tech, February 13, 2015, http://money.cnn.com/2015/02/12/technology/living-without-cell-phone/index.html.

117 **"Only thoughts reached by walking":** Friedrich Nietzsche. *Twilight of the Idols* (1889), maxim 34, http://www.lexido.com/ebook_texts/twilight_of_the_idols_.aspx?S=2.

117 **"The sedentary life"**: Nietzsche, *Twilight*, maxim 34.

117 **"he became the peerless walker"**: Gros, *A Philosophy of Walking*, 16.

118 **the example of the French poet Arthur Rimbaud**: Gros, *A Philosophy of Walking*, 39–47.

118 **"I never do anything but"**: Jean-Jacques Rousseau, as quoted in Gros, *A Philosophy of Walking*, 65.

118 **"The mere sight of a desk"**: Gros, *A Philosophy of Walking*, 65.

118 **"As I walk, I am always reminded"**: Wendell Berry, "Wendell Berry: The Work of Local Culture," *The Contrary Farmer: Gene Logsdon Memorial Blogsite*, June 10, 2011, https://thecontraryfarmer.wordpress.com/2011/06/10/wendell-berry-the-work-of-local-culture.

118 **"The walking of which I speak"**: Henry David Thoreau, "Walking," *Atlantic Monthly*, June 1862, https://www.theatlantic.com/magazine/archive/1862/06/walk ing/304674.

119 **"We do not belong"**: quoted in Gros, *A Philosophy of Walking*, 18.

122 **"I think that I cannot preserve"**: Thoreau, "Walking."

126 **Dwight Eisenhower leveraged**: Kethledge and Erwin, *Lead Yourself First*, 35.

126 **Abraham Lincoln had a habit**: Cali, "Lincoln's Emancipation Proclamation."

CHAPTER 5: DON'T CLICK "LIKE"

127 **ESPN aired what has to be one of the strangest sporting events**: "2007 USARPS Title Match," YouTube video, 3:58, recorded July 7, 2007, posted by "usarpsleague," October 8, 2007, https://www.youtube.com/watch?v=_eanWnL3FtM.

128 **the role of skill becomes**: For more on the claim that high-ranked players perform consistently better than what would be expected if the game's outcome were random, see Alex Mayyasi, "Inside the World of Professional Rock Paper Scissors," *Priceonomics*, April 26, 2016, https://priceonomics.com/the-world-of-competitive-rock-paper-scissors.

128 **In a promotional video**: "Street rps," YouTube video, 1:24, posted by "usrpsleague," January 18, 2009, https://www.youtube.com/watch?v=6QWPbi3-nlc.

131 **"man is by nature a social animal"**: Aristotle, *Politics: Books I., III., IV. (VII.)*, trans. W. E. Bolland (London: Longmans, Green, and Co., 1877), 112.

131 **published a pair of papers**: Gordon L. Shulman, Maurizio Corbetta, Randy Lee Buckner, Julie A. Fiez, Francis M. Miezin, Marcus E. Raichle, and Steven E. Petersen, "Common Blood Flow Changes across Visual Tasks: I. Increases in Subcortical Structures and Cerebellum but Not in

Nonvisual Cortex," *Journal of Cognitive Neuroscience* 9, no. 5 (October 1997): 624–47, https://doi.org/10.1162/jocn.1997.9.5.624; Gordon L. Shulman, Julie A. Fiez, Maurizio Corbetta, Randy L. Buckner, Francis M. Miezin, Marcus E. Raichle, and Steven E. Petersen, "Common Blood Flow Changes across Visual Tasks: II. Decreases in Cerebral Cortex," *Journal of Cognitive Neuroscience* 9, no. 5 (October 1997): 648–63, doi:10.1162 /jocn.1997.9.5.648.

131 **"only a few regions showed increased activity":** Matthew D. Lieberman, *Social: Why Our Brains Are Wired to Connect* (New York: Crown, 2013), 16.

132 **"it was an unusual question":** Lieberman, *Social*, 16.

133 **"other people, yourself, or both":** Lieberman, *Social*, 18.

133 **"virtually identical":** Lieberman, *Social*, 18.

133 **"I have since become convinced":** Lieberman, *Social*, 19.

134 **"clearly haven't cultivated an interest":** Lieberman, *Social*, 20.

134 **"The brain did not evolve":** Lieberman, *Social*, 15.

136 **The first was an NPR story posted in March:** Katherine Hobson, "Feeling Lonely?: Too Much Time on Social Media May Be Why," NPR, March 6, 2017, https://www.npr.org/sections/health-shots/2017/03/06 /518362255/feeling-lonely-too-much-time-on-social-media-may-be-why.

137 **"when used properly":** David Ginsberg and Moira Burke, "Hard Questions: Is Spending Time on Social Media Bad for Us?," Newsroom, Facebook, December 15, 2017, https://newsroom.fb.com/news/2017/12/hard -questions-is-spending-time-on-social-media-bad-for-us.

137 **"brings us joy and strengthens":** Ginsberg and Burke, "Spending Time on Social."

137 **One of the main positive articles cited by the Facebook blog post:** Moira Burke and Robert E. Kraut, "The Relationship Between Facebook Use and Well-Being Depends on Communication Type and Tie Strength," *Journal of Computer Mediated Communication* 21, no. 4 (July 2016): 265–81, https://doi.org/10.1111/jcc4.12162.

138 **Another positive article cited in the Facebook post:** Fenne große Deters and Matthias R. Mehl, "Does Posting Facebook Status Updates Increase or Decrease Loneliness? An Online Social Networking Experiment," *Social Psychological and Personality Science* 4, no. 5 (September 2013): 579–86, https://doi.org/10.1177/1948550612469233.

139 **The first of these studies was authored by a large team:** Brian A. Primack, Ariel Shensa, Jaime E. Sidani, Erin O. Whaite, Liu yi Lin, Daniel Rosen, Jason B. Colditz, Ana Radovic, and Elizabeth Miller, "Social Media Use and Perceived Social Isolation among Young Adults in the U.S.," *American Journal of Preventive Medicine* 53, no. 1 (July 2017): 1–8, https://doi.org/10.1016/j.amepre.2017.01.010.

139 **"It's social media, so aren't people"**: Hobson, "Feeling Lonely?"

140 **"Our results show that overall"**: Holly B. Shakya and Nicholas A. Christakis, "Association of Facebook Use with Compromised Well-Being: A Longitudinal Study," *American Journal of Epidemiology* 185, no. 3 (February 2017): 203–11, https://doi.org/10.1093/aje/kww189.

140 **These negative connections still held**: Shakya and Christakis, "Association of Facebook Use," 205–6.

141 **"What we know at this point"**: Hobson, "Feeling Lonely?"

142 **"Where we want to be cautious"**: Hobson, "Feeling Lonely?"

144 **"Face-to-face conversation is the most human"**: Sherry Turkle, *Reclaiming Conversation: The Power of Talk in a Digital Age*, rev. ed. (New York: Penguin Books, 2016), 3.

145 **"flight from conversation"**: Turkle, *Reclaiming Conversation*, 4.

145 **"Don't all these little tweets"**: Turkle, *Reclaiming Conversation*, 34. The appearance on *The Colbert Report* that Turkle described in this cited passage from *Reclaiming Conversation* originally aired on January 17, 2011.

145 **"Face-to-face conversation unfolds slowly"**: Turkle, *Reclaiming Conversation*, 35.

146 **"My argument is not anti-technology"**: Turkle, *Reclaiming Conversation*, 25.

146 **"seriousness of the moment"**: Turkle, *Reclaiming Conversation*, 4.

150 **just five days at a camp with no phones or internet**: Turkle, *Reclaiming Conversation*, 11.

151 **Facebook didn't invent the "Like" button**: "What's the History of the Awesome Button (That Eventually Became the Like Button) on Facebook?," Quora, answer by Andrew "Boz" Bosworth, updated October 16, 2014, https://www.quora.com/Whats-the-history-of-the-Awesome-Button -that-eventually-became-the-Like-button-on-Facebook.

152 **As Chan explains, many Facebook posts**: Kathy H. Chan, "I Like This," Notes, Facebook, February 9, 2009, https://www.facebook.com /notes/facebook/i-like-this/53024537130.

155 **"I don't think we're meant"**: Jennifer Grygiel, assistant professor, S.I. Newhouse School of Public Communication, phone interview with author, January 26, 2018.

156 **"Phones have become woven into"**: Turkle, *Reclaiming Conversation*, 158.

160 **Sherry Turkle calls this effect "phone phobia"**: Turkle, *Reclaiming Conversation*, 148.

CHAPTER 6: RECLAIM LEISURE

165 **"The best and most pleasant life"**: Aristotle, *Ethics*, trans. J. A. K. Thomson, rev. ed. (New York: Penguin Books, 2004), 273.

165 **"activity that is appreciated"**: Aristotle, *Ethics*, 271.
166 **"worth depends on the existence"**: Kieran Setiya, *Midlife: A Philosophical Guide* (Princeton, NJ: Princeton University Press, 2017), 43.
166 **"source of inward joy"**: Setiya takes the phrase "source of inward joy" from John Stuart Mill's self-reported recovery from depression by finding beauty in poetry—an activity he could pursue purely for the sake of its beauty. See Setiya, *Midlife*, 40, 45.
167 **"By the end of day two"**: Harris, *Solitude*, 220.
167 **"I remember that this"**: Harris, *Solitude*, 219.
171 **"I never understood the joy"**: "Seek Not to Be Entertained," *Mr. Money Mustache* (blog), September 20, 2017, https://www.mrmoneymustache.com/2017/09/20/seek-not-to-be-entertained.
172 **Mr. Money Mustache World Headquarters:** "Introducing The MMM World Headquarters Building," *Mr. Money Mustache* (blog), August 2, 2017, http://www.mrmoneymustache.com/2017/08/02/introducing-the-mmm-world-headquarters-building.
172 **"If you leave me alone"**: "Seek Not," *Mr. Money Mustache*.
172 **"even if it's ten below outside"**: Liz Thames, phone interview by the author, December 20, 2017.
173 **"For me, inactivity leads"**: "Seek Not," *Mr. Money Mustache*.
174 **"I wish to preach"**: Theodore Roosevelt, "The Strenuous Life" (speech before the Hamilton Club, April 10, 1899), http://www.bartleby.com/58/1.html.
175 **"acquainted with a genuinely good whiskey"**: Arnold Bennett, *How to Live on 24 Hours a Day* (New York: WM. H. Wise & Co., 1910), 37.
175 **"gone like magic"**: Bennett, *How to Live*, 37.
175 **"never demand any appreciable mental application"**: Bennett, *How to Live*, 66.
175 **"mental strain"**: Bennett, *How to Live*, 67.
176 **"What? You say that full energy"**: Bennett, *How to Live*, 32–33.
178 **"People have the need"**: Gary Rogowski, *Handmade: Creative Focus in the Age of Distraction* (Fresno: Linden Publishing, 2017), 157.
178 **"Long ago we learned to think"**: Rogowski, *Handmade*, 156.
178 **"Many people experience the world"**: Rogowski, *Handmade*, 156.
179 **"They seem to relieve him"**: Matthew B. Crawford, "Shop Class as Soulcraft," *New Atlantis*, no. 13 (Summer 2006): 7–24, https://www.thenewatlantis.com/publications/shop-class-as-soulcraft.
182 **"Leave good evidence of yourself"**: Rogowski, *Handmade*, 177.
183 **best exemplified by the megahit Settlers of Catan:** Dave McNary, "'Settlers of Catan' Movie, TV Project in the Works," *Variety*, February 19, 2015, https://variety.com/2015/film/news/settlers-of-catan-movie-tv-project-gail-katz-1201437121.

183 **"Tabletop gaming creates a unique social space"**: David Sax, *The Revenge of Analog: Real Things and Why They Matter*, trade paperback ed. (New York: PublicAffairs, 2017), 80.

183 **"a rich multimedia, 3D interaction"**: Sax, *Revenge of Analog*, 82.

183 **"the signal flares of our most complex emotions"**: Sax, *Revenge of Analog*, 83.

184 **"On a social level, video games"**: Sax, *Revenge of Analog*, 83.

185 **"fitness has shifted from a private activity"**: Matt Powell, "Sneakernomics: How 'Social Fitness' Changed the Sports Industry," *Forbes*, February 3, 2016, https://www.forbes.com/sites/mattpowell/2016/02/03/sneakernomics-how-social-fitness-changed-the-sports-industry.

186 **The site then provides a "lexicon" of F3 jargon**: "Lexicon," F3, http://f3nation.com/lexicon, accessed July 14, 2018.

187 **there are over 1,200 groups**: "Where Is F3," F3, https://f3nation.com/workouts, accessed July 14, 2018.

187 **there's one CrossFit box for every two Starbucks**: "Find a Box," CrossFit, https://map.crossfit.com/; "Number of Starbucks Stores Worldwide from 2003 to 2017," Statista, https://www.statista.com/statistics/266465/number-of-starbucks-stores-worldwide/; Christine Wang, "How a Health Nut Created the World's Biggest Fitness Trend," CNBC, April 5, 2016, https://www.cnbc.com/2016/04/05/how-crossfit-rode-a-single-issue-to-world-fitness-domination.html.

188 **Here's a sample WOD**: "Friday 171229," Workout of the Day, CrossFit, https://www.crossfit.com/workout/2017/12/29#/comments.

188 **"The camaraderie of other members"**: Steven Kuhn, "The Culture of CrossFit: A Lifestyle Prescription for Optimal Health and Fitness" (senior thesis, Illinois State University, 2013), 12, https://ir.library.illinoisstate.edu/cgi/viewcontent.cgi?article=1004&context=sta.

189 **"religion run by a biker gang"**: Glassman has called CrossFit a "religion run by a biker gang" on many public occasions, e.g., Catherine Clifford, "How Turning CrossFit into a Religion Made Its Atheist Founder Greg Glassman Rich," CNBC, October 11, 2016, https://www.cnbc.com/2016/10/11/how-turning-crossfit-into-a-religion-made-its-founder-atheist-greg-glassman-rich.html.

190 **The Mouse Book Club provides a good example**: For more on the Mouse Book Club, see https://mousebookclub.com.

190 **"mobilizing literature"**: "About," Mouse Books Kickstarter campaign, https://www.kickstarter.com/projects/mousebooks/mouse-books.

194 **"Damn! . . . If this guy is billing out"**: "Unlock Your Inner Mr. T—by Mastering Metal," *Mr. Money Mustache* (blog), April 16, 2012, http://www.mrmoneymustache.com/2012/04/16/unlock-your-inner-mr-t-by-mastering-metal.

195–96 **"It was simply taken for granted"**: Crawford, "Soulcraft."

199 **"I just don't appreciate social networking"**: "Jim Clark in Conversation with John Hennessey," YouTube video, 1:04:07, recorded May 23, 2013, posted by "stanfordonline," June 26, 2013, https://www.youtube.com /watch?v=gXuOH9B6kTM.

199 **"[The executive was] just raving"**: "Jim Clark in Conversation," YouTube.

203 **"I had form'd most"**: Benjamin Franklin, *The Autobiography of Benjamin Franklin* (New York, 1909; Project Gutenberg, 1995), pt. 1, http://www .gutenberg.org/files/148/148-h/148-h.htm.

CHAPTER 7: JOIN THE ATTENTION RESISTANCE

213 **"digital technologies transform how we live"**: Elliot Schrage, "Introducing Hard Questions," Newsroom, Facebook, June 15, 2017, https:// newsroom.fb.com/news/2017/06/hard-questions.

214 **"Society often welcomes the benefit"**: Rob Sherman, "Hard Questions: Should I Be Afraid of Face Recognition Technology?," Newsroom, Facebook, December 19, 2017, https://newsroom.fb.com/news/2017/12 /hard-questions-should-i-be-afraid-of-face-recognition-technology.

214 **"a lot of smart people"**: Ginsberg and Burke, "Spending Time on Social."

214 **"According to the research"**: Ginsberg and Burke, "Spending Time on Social."

215 the **"attention economy" describes the business sector**: On "attention economy," see Tim Wu, *The Attention Merchants: The Epic Scramble to Get Inside Our Heads* (New York: Alfred A. Knopf, 2016).

215 **beginning of this economic model to 1830**: Tim Wu, "The Battle for Our Attention," October 25, 2016, Shorenstein Center, Harvard University, transcript highlights and Soundcloud audio, 1:04:04, https://shoren steincenter.org/tim-wu.

215 **"He was the first person"**: Wu, "Battle for Our Attention."

216 **The most expensive internet company at the time was eBay**: Alex Wilhelm, "A Look Back in IPO: Google, the Profit Machine," *Tech-Crunch*, July 31, 2017, https://techcrunch.com/2017/07/31/a-look-back-in -ipo-google-the-profit-machine.

216 **Google is the second most valuable company in the United States:** "U.S. Commerce—Stock Market Capitalization of the 50 Largest American Companies," iWeblists, accessed January 31, 2018, http://www.iweb lists.com/us/commerce/MarketCapitalization.html.

217 **fifty minutes *per day* on Facebook products alone**: David Cohen, "How Much Time Will the Average Person Spend on Social Media

During Their Life? (Infographic)," *Adweek*, March 22, 2017, http://www
.adweek.com/digital/mediakix-time-spent-social-media-infographic.

220 **"[Twitter] scares me"**: George Packer, "Stop the World," *New Yorker*,
January 29, 2010, https://www.newyorker.com/news/george-packer/stop
-the-world.

222 **By October, 14 percent of the company's ad revenue:** Josh Constine,
"Study: 20% of Ad Spend on Facebook Now Goes to Mobile Ads," *Tech-
Crunch*, January 7, 2013, https://techcrunch.com/2013/01/07/facebook
-mobile-ad-spend.

222 **Facebook reported that 62 percent of its revenue:** Ellis Hamburger,
"Facebook's New Stats," *The Verge*, July 23, 2014, https://www.theverge
.com/2014/7/23/5930743/facebooks-new-stats-1-32-billion-users
-per-month-30-percent-only-use-it-on-their-phones.

222 **by 2017, mobile ad revenue rose to 88 percent:** "Ad Revenue Growth
Continues to Propel Facebook," *Great Speculations* (blog), *Forbes*, November 2, 2017, https://www.forbes.com/sites/greatspeculations/2017/11/02
/ad-revenue-growth-continues-to-propel-facebook/#54b22b2865ed.

222 **mobile pays the bills:** For a more detailed breakdown of Facebook revenue, see this summary on their website of the most recent quarterly report
(at the time of this writing), which now has mobile ad revenue at 89
percent: https://investor.fb.com/investor-news/press-release-details/2018
/Facebook-Reports-Fourth-Quarter-and-Full-Year-2017-Results
/default.aspx.

225 **He called it, appropriately enough, Freedom:** For more on the Freedom software, its features, its user count, and research on increased productivity, see https://freedom.to/about.

226 **the novelist Zadie Smith:** Vijaysree Venkatraman, "Freedom Isn't
Free," *Science*, February 1, 2013, http://www.sciencemag.org/careers/2013
/02/freedom-isnt-free.

226 **"There's an even deeper irony":** Venkatraman, "Freedom Isn't Free."

227 **IBM was selling automatic tabulating machines:** For more on IBM's
early history, see http://www-03.ibm.com/ibm/history/history/year_1890
.html. Note that IBM did not take on the name International Business
Machines until 1924.

227 **An early print ad for the Apple II:** Buster Hein, "12 of the Best Apple
Print Ads of All Time (Gallery)," Cult of Mac, October 17, 2012, https://
www.cultofmac.com/196454/12-of-the-best-apple-print-ads-of-all
-time-gallery.

232 **"If you [look at my Twitter feed,] you won't see a lot of dog meme
accounts":** Jennifer Grygiel, assistant professor, S.I. Newhouse School of
Public Communication, phone interview with author, January 26, 2018.

236 **Das Slow Media Manifest:** Das Slow Media Manifest, Slow Media Institut, http://slow-media-institut.net/manifest.
236 **The English translation:** The Slow Media Manifesto, English translation, Slow Media Institute, http://en.slow-media.net/manifesto.
236 **"brought profound changes":** Slow Media Manifesto.
236 **"appropriate reaction":** Slow Media Manifesto.
236 **embrace the concept of "slow":** Slow Media Manifesto.
237 **"Slow media cannot be consumed":** Slow Media Manifesto.
237 **Americans tend to embrace the "low information diet":** Timothy Ferriss first popularized the term "low information diet" in *The 4-Hour Workweek: Escape 9–5, Live Anywhere, and Join the New Rich* (New York: Crown, 2007).
243 **"It's silly, I know, but the first few weeks felt rough":** Quotes from Paul come from email correspondence with the author that was conducted primarily in December 2015.
244 **"I feel so much better":** Daniel Clough, "Feature Phones Aren't Just for Hipsters," November 20, 2015, http://danielclough.com/feature-phones-arent-just-for-hipsters.
244 **"is not as drastic a regression":** Vlad Savov, "It's Time to Bring Back the Dumb Phone," *The Verge*, January 31, 2017, https://www.theverge.com/2017/1/31/14450710/bring-back-the-dumb-phone.
244–45 **These products, which include:** For details on the Light Phone, see https://www.thelightphone.com.
245 **"Quickly it became obvious":** "About," Light Phone, https://www.thelightphone.com/about.
246 **"Your [clock symbol]":** "About," Light Phone.

CONCLUSION

249 **"experienced the epiphany":** Simon Winchester, *The Men Who United the States: America's Explorers, Inventors, Eccentrics, and Mavericks, and the Creation of One Nation, Indivisible* (New York: HarperCollins, 2013), 338. For the reader interested in a detailed account of the telegraph's invention and subsequent impact, see Winchester, *The Men*, 335–57; Tom Standage, *The Victorian Internet: The Remarkable Story of the Telegraph and the Nineteenth Century's On-Line Pioneers* (New York: Walker & Co., 1998).
249 **"If the presence of electricity":** Winchester, *The Men*, 339.
249 **"vatic revelation":** Winchester, *The Men*, 339.
250 **"formed a simple declarative exclamation":** Winchester, *The Men*, 347.
251 **"so mystifyingly different":** Winchester, *The Men*, 336.

Index

PENGUIN PARTNERSHIPS

Penguin Partnerships is the Creative Sales and Promotions team at Penguin Random House. We have a long history of working with clients on a wide variety of briefs, specializing in brand promotions, bespoke publishing and retail exclusives, plus corporate, entertainment and media partnerships.

We can respond quickly to briefs and specialize in repurposing books and content for sales promotions, for use as incentives and retail exclusives as well as creating content for new books in collaboration with our partners as part of branded book relationships.

Equally if you'd simply like to buy a bulk quantity of one of our existing books at a special discount, we can help with that too. Our books can make excellent corporate or employee gifts.

Special editions, including personalized covers, excerpts of existing books or books with corporate logos can be created in large quantities for special needs.

We can work within your budget to deliver whatever you want, however you want it.

For more information, please contact
salesenquiries@penguinrandomhouse.co.uk